# FLEXIBLE PACKAGING

# OF FOODS

author:

AARON L. BRODY
Arthur D. Little, Inc.
Food and Flavor Section
Cambridge, Mass.

CRC Press
Taylor & Francis Group
Boca Raton  London  New York

CRC Press is an imprint of the
Taylor & Francis Group, an **informa** business

# CRC MONOSCIENCE SERIES

The primary objective of the CRC Monoscience Series is to provide reference works, each of which represents an authoritative and comprehensive summary of the "state-of-the-art" of a single well-defined scientific subject.

Among the criteria utilized for the selection of the subject are: (1) timeliness; (2) significant recent work within the area of the subject; and (3) recognized need of the scientific community for a critical synthesis and summary of the "state-of-the-art."

The value and authenticity of the contents are assured by utilizing the following carefully structured procedure to produce the final manuscript:

1. The topic is selected and defined by an editor and advisory board, each of whom is a recognized expert in the discipline.

2. The author, appointed by the editor, is an outstanding authority on the particular topic which is the subject of the publication.

3. The author, utilizing his expertise within the specialized field, selects for critical review the most significant papers of recent publication and provides a synthesis and summary of the "state-of-the-art."

4. The author's manuscript is critically reviewed by a referee who is acknowledged to be equal in expertise in the specialty which is the subject of the work.

5. The editor is charged with the responsibility for final review and approval of the manuscript.

In establishing this new CRC Monoscience Series, CRC has the additional objective of attacking the high cost of publishing in general, and scientific publishing in particular. By confining the contents of each book to an *in-depth treatment* of a relatively narrow and well-defined subject, the physical size of the book, itself, permits a pricing policy substantially below current levels for scientific publishing.

Although well-known as a publisher, CRC now prefers to identify its function in this area as the management and distribution of scientific information, utilizing a variety of formats and media ranging from the conventional printed page to computerized data bases. Within the scope of this framework, the CRC Monoscience Series represents a significant element in the total CRC scientific information service.

B. J. Starkoff, President
THE CHEMICAL RUBBER Co.

This book originally appeared as part of an article in *CRC Critical Reviews in Food Technology,* a quarterly journal published by The Chemical Rubber Co. We would like to acknowledge the editorial assistance received by the Journal's editor, Thomas E. Furia, Geigy Industrial Chemicals. Mr. Stanley Sacharow, Reynolds Metals Co., served as referee for this article.

# AUTHOR'S INTRODUCTION

Packaging is an integral element of the system that brings food to American consumers. Although packaging has sometimes been viewed as an inconsequential act of inserting food into a film or box, this attitude has been a prime source of product failure. In self-service retailing, the package is the sole communication link between processor and consumer at the most critical moment of a product's life—that instant the consumer makes the purchase decision. The package telegraphs to the consumer the identity of the product and its processor, all in a fleeting moment. The package continues its communication role as the major consumer advertising message on the kitchen shelf. These key marketing functions must be performed while the package is protecting the product against the environment.

Package and product must be combined with efficiency and speed to assure quality retention in the product entering the distribution system. And these functions are performed at a cost that has enabled American manufacturers to sell food at the lowest price in history and in the world.

Effective performance of this multitude of functions requires blending of a large number of divergent disciplines that, together, are packaging. Design and graphics; materials and mechanics; chemistry and physics are but a sampling of the many components that constitute packaging. To study packaging as just material or machine specifications is to fail to integrate the needs of the product and its market with packaging.

Effective packaging begins with an understanding of the requirements of the product and its marketing. These needs can then be logically married to materials and machines to create a packaging system.

Food technologists and marketers should perform better with knowledge of the packaging requirements of food products and of how industry has met these requirements. A study of a broad range of food product requirements and packaging can be a stimulus to more profitably applying good packaging to food.

This review encompasses one segment of food packaging—foods that include flexible materials in their package. It is intended as a systems viewpoint: defining product requirements and markets, and describing the present material and machine methods for filling those requirements.

*Flexible Packaging of Foods* and its underlying philosophies are the result of the thinking and teaching of many people. The author expresses his appreciation to the packaging suppliers and food technologists who contributed to his understanding over the years but who could not possibly be listed. This publication could not have been possible without the presence of some who merit very special thanks— Mrs. Janet Higgins and Miss Anne Lansky, secretaries who made the text readable; Mr. Loren B. Sjöström, Vice President, Food and Flavor Section, Arthur D. Little, Inc., who stimulated and encouraged the project; Stephen, Glen, and Robyn, who were so quiet and patient while Daddy worked; and, most important of all, my wonderful wife, Carolyn.

## THE AUTHOR

**Aaron L. Brody** is a staff member of the Food and Flavor Section of Arthur D. Little, Inc. Dr. Brody earned his S.B. and Ph.D. degree from Massachusetts Institute of Technology and an M.B.A. from Northeastern University. He has authored numerous articles on packaging.

# TABLE OF CONTENTS

By any measurement, packaging is a large industry in size, diversity and complexity. Mere enumeration of the amount of materials consumed, or dollar volume of machinery sold, would fail to depict the numbers of persons who package products each day. Entire industries are in reality not manufacturers but packagers of products made by or assembled by others. Almost all goods reaching the industrial or consumer level are packaged. To ascribe a dollar volume in excess of $25 billion to packaging would be to miss much of the value.

Clear segmentation of so large a group of industries is difficult. Packaging might be divided into cans, glass, paper, paperboard, plastic and flexible. This classification, based on materials, overlaps and lacks clarity. Most paper and some plastic packaging is flexible. A rigid plastic tub with a laminate lid is a rigid container, but it might perhaps have greater flexible value than rigid material. Aluminum foil laminated to paper is used as label stock on cans and glass, but it is generally not considered to be flexible packaging.

A precise definition of flexible packaging is not important. To assure that the significant uses of flexible packaging are examined, a number of areas which might be marginal are covered. Flexible packaging, then, includes a broad range of papers, plastics, aluminum foils, coatings, bondings, and various combinations which generally are not self-supporting in flat form and which can be shaped.

Estimates have placed the volume of flexible packaging used annually in recent years at 1.6 billion pounds valued at $1.4 billion. The growth rate has been estimated at 7% per year.

Although multi-wall bags are usually not considered flexible packages, a large volume of material is employed to contain bulk materials such as fertilizers and chemicals. Similarly, relatively small quantities of expensive and sophisticated flexible materials are used by the pharmaceutical, cosmetic and toiletries industry to protect their products. The tobacco industry uses large quantities of flexible materials, including cellophane, paper, and aluminum foil laminations to keep products fresh. Clothing is wrapped in considerable quantities of flexible packaging. Durable consumer goods, such as phonograph records and toys, employ flexible materials both for protection and decoration.

However, the largest single use category is probably food. The costs of raw materials and commodities for food products, coupled with a fierce competitive situation among processors, has led to generally low profit margins among food industries. Efforts to maintain or improve profitability by food companies have ranged from mechanization to reduction of labor.

Flexible packaging has benefited from this pressure for lower cost. Much food packaging was in rigid containers which have persisted. Rigid glass and metal are protective, impermeable and, when properly coated and closed, inert to the contents. Further, rigid glass and metal are able to withstand the stresses of thermal or freezing processing. Because of their structure, rigid containers lend themselves to high-speed automatic equipment. Rigid containers are, however, among the highest cost packaging used for foods. The fact that rigid containers are difficult to fabricate has mitigated against the food industry manufacturing, and dictated that these containers be fully formed at the packaging suppliers' plants. The volume of fabricated glass and metal containers used requires large warehouse facilities or skillful logistics to assure adequate supplies. Although some food processors have their own facilities for can manufacturing, cans and glass demand long run quantities for efficient operation and lower cost.

All of these factors have worked in favor of flexible materials which are inherently cheaper than rigid materials. Flexible packaging is also flexible from a business standpoint. Although efficiencies can be realized from long runs, even short runs are not difficult to manufacture at moderate cost. A broad range of properties may be built into flexible packaging with appropriate components. Size, for example, can be varied, sometimes at the packaging site. And, since flexible materials are made in the flat, they do not require large inventory facilities.

Flexible materials may be obtained in preformed configuration which is filled and closed by the food packager, or in roll or sheet form. The food packager may manufacture the package in-plant from roll or sheet stock, thus eliminating part of the price of vendor package fabrication. A considerable amount of this package fabrication equipment forms the package around the product or forms the package and fills it in sequence. Such machinery allows for integration of product manufacture directly into the package and thus reduces costs.

There may be some sacrifice of structural strength, inertness and permeability which limits the applications of flexible packaging to foods. For example, no known coating is as inert as glass. Being lighter in weight, flexible materials might be thinner and weaker than rigid.

Because of the multitude of basic paper, plastic and ink suppliers, and the similarity of flexible packaging converting to the printing trade, there are many more flexible packaging vendors than suppliers of rigid containers. This leads to a lesser dependence on a few sources of supply. Many converters provide a specialized line of flexible materials to a limited part of the food packaging industries. Some converters have direct interest in raw materials such as a plastic or paper. No converter, however, has infinite capability in a field as complex as flexible packaging of foods.

The field of flexible packaging of foods is dynamic. Nevertheless, however accurate a picture of an existing situation is, the picture is transient. At present, the relatively limited number of commercial plastics, papers, foils, coatings and adhesion systems give rise to an estimated seven million possible combinations of flexible materials.

Polymer chemists have developed many more materials which are not commercial; however, applications have been detailed for these materials. Large research efforts have been directed toward coatings to change the properties of the basic films. Even paper has been the subject of intensive research. Basic material suppliers are seeking new means for upgrading their output and for widening potential applications. Working in cooperation with their customers (the converters) and the ultimate customers (the packagers), basic material suppliers are constantly creating new combinations of flexible materials. Converters, caught between basic material suppliers and customers striving for lower cost and higher performance materials, are developing packaging to better their position.

The packagers, in their quest for new products and better means for presenting them, are making increasing demands upon their suppliers. These statements of customer need are leading the suppliers to develop new flexible packaging.

Equipment manufacturers, most often small and highly specialized, create part of the pressure which leads to new flexible materials. Concepts of higher speed and more efficient packaging machinery have directed converter and basic supplier attention to flexible materials suitable for the equipment designs. The need for assuring the proper marriage of material to machine has led to increases in joint ventures and in organizations which operate in both areas.

Two areas of government have played a prominent role in package specification. Safety, by virtue of materials in contact with foods, is the province of Food and Drug Administration. Also under FDA jurisdiction are consumer protection and communication elements, such as assurance of fill and adequacy of information on the label. By its nature, flexible packaging may not appear to contain the declared quantity of contents if filled on high speed automatic equipment. Government has also instituted a program of package shape and size standardization in an effort to help the consumer shop without a runaway proliferation of combinations and permutations of sizes and weights for similar products.

With these many and diverse resources lending themselves to problems of flexible packaging, improvements and solutions are regular inputs to food packagers. In the new product area, new packaging systems are the rule.

Certain fundamentals, such as requirements for containment, protection and decoration, remain. Several of the original films have survived as basic materials to flexible packaging. These include papers, cellophanes, polyethylenes and aluminum foils. All have been improved for specific uses. Other materials have entered the market, and, of course, more will become part of the packaging world with the passage of time.

Both basic material suppliers and converters constantly review the flexible packaging situation. Their useful documents are proprietary to those who have prepared them, and focus has been on the specific and obviously relatively narrow interests of the companies preparing these reports. Users have occasionally been privileged to study portions of these reports relative to the immediate and direct product interests of the users.

Examination of a detailed review would present opportunities for cross-fertilization not generally afforded users. Presentation of the totality of flexible packaging of foods in one place for users would offer both users and their suppliers perspectives which might clarify certain problems and define solutions based on experiences in other segments of food packaging. Presentation from the standpoint of the food, rather than from the packaging material, might offer a view that has not been afforded either the user or supplier. Even though this field is constantly changing, there is purpose to documenting the situation at a single point of time.

As implied, flexible packaging industries include basic material suppliers, converters, machinery manufacturers and packagers. In general, basic material suppliers manufacture paper, plastics, inks, coatings and metal foils. Converters are those who make the basic materials useful by coating, laminating, printing and cutting. Machinery makers design and build equipment which transforms the converted flexible materials into filled packages. The packagers are those who bring product and packaging material together and produce the package.

Generally, science is found among the basic material producers. Applications orientation is most often with the packager. Converters are generally concerned with building properties into materials. The converter may be considered a middle man, attempting to use what is available to satisfy the needs of the users. Thus, the converters' problems may be greater than those of the industries which surround them. The magnitude of converter responses to the pressures exerted from two sides has often restricted converter development to immediate problem solving. Only a few large converters have the resources for long-range development.

As a result of this natural division of interests, the limited number of technical and trade publications are generally devoted to materials, their components, properties and manufacturing. Similarly, machinery designers' literature focuses upon the mechanical characteristics of the equipment. The many basic material producers' technical publications describe many technical properties of the chemical compounds.

The packagers themselves are usually concerned with the results of packaging and often have little knowledge of the identity of the material/machine combinations used for their products. Technical publications on food products usually treat packaging in a most cursory fashion, if at all. Product technologists often leave packaging to purchasing or marketing people.

This paucity of published information on technical interactions of product and packaging has limited commercialization of some important packaging developments. For example, perhaps the greatest number of technical publications on flexible packaging deal with fresh red meat. Most students of this topic concede that despite the many publications there are enormous voids of information. Some of the needed data is believed to have been prepared by various suppliers, but little of this has been integrated for users. One reason for this lack of dissemination is that those who possess needed information may not know the potential market.

This background demonstrates the lack of a comprehensive source of applications information on flexible packaging. Such a review would not be an instruction manual on flexible packaging analogous to those on canning, paperboard packaging equipment or corrugated case design. Rather, this review is aimed at describing food products employing flexible packaging, the requirements dictating the flexible packaging being used and the current state of flexible packaging in the food industry. Materials and their properties are included as needed, but emphasis is given to stated and inferred marketing needs which influence the selection of materials and their graphics. Machinery is indicated in terms of describing the packaging that it makes and the rationale for including it in a production line.

It is expected that this attempt to cover a vast field will lead to some omissions, but perhaps the inclusion of a broad range will lead to later insertion of the missing parts.

Because of the dynamics of flexible food packaging, some of this review will be obsolete by the time this article is printed. In these areas, this document serves as an historic record and thus might still interest and stimulate the persons with direct packaging responsibilities.

In few other reviews would it be so difficult to classify the subject areas. In an attempt to minimize reader difficulty in reviewing and referencing, this report is classified in alphabetic order according to food categories:

Bakery
    Bread, cakes, soft goods, pies
Biscuit and cracker
    Cookies and crackers
Canned foods
    Sterilized flexible packaging
    Film multipacking
Cereals
    Ready-to-eat and hot breakfast
Coffee
    Roasted and ground, and instant
Confectionery
    Candy and nuts
Convenience foods
    Food service and retail
Dairy
    Fluid milk, ice cream, dry milk, cheese, butter and margarine
Dehydrated mixes
    Formulated mixes, cake mixes, dinner mixes, rice, flour, pasta, soups, salad dressing and sauce mixes.
Desserts
    Mixes and refrigerated
Frozen foods
    Fresh and precooked
Meat
    Red and processed
Poultry
Produce
Snacks

For each product category, the products included, their manufacture, distribution and marketing are briefly outlined. From the marketing or ultimate user standpoint, the packaging requirements are reviewed, pointing out the important needs served by packaging. Present packaging for each product group within the classification is described, with the detail on flexible packaging. There may be omissions from the descriptions because information is

FIGURE 1

Definitions for Some Flexible Packaging Materials Used for Foods

## Flexible-packaging terminology — Definitions proposed by MODERN PACKAGING

| Definition | Notes |
|---|---|
| **Film:** A non-fibrous, basically organic, thin material made by extruding, casting, calendering or regenerating polymers, plastic compounds or cellulose in thicknesses not exceeding 3 mils. | 1. Films generally are used in pouching, bagging, wrapping, shrink- and skin-packaging operations. 2. Defining "film" by stiffness or function is impractical. However, as some plastics more than 3 mils thick are used within the accepted scope of flexible packaging, these will be known as "heavy-duty films" when less than 10 mils thick. Thicker materials are designated "sheet." 3. "Film" does not apply to webs other than plastics and cellophane, but see Note 3 immediately below. |
| **Basic film:** A film that is homogeneous in cross-section and therefore not a laminated or coated web. | 1. Basic films are composed of comonomers, copolymers, ionomers, similar chemical substances and blends of these ingredients. 2. "Basic film" replaces such terms as "free" and "unsupported." These terms continue to have other specific connotations, i.e., "free-film" plies in multiwall bags. 3. Cellophanes used in packaging almost always are not basic films because they are coated. "Structured web" or "structured" or "coated film" are correct terms. |
| **Structured web:** A web that is not homogeneous in cross-section and is composed of two or more fundamentally similar materials. | 1. All layers must be only films or only papers or only foil—"fundamentally similar" materials. Such webs would also be known, respectively, as "structured film," "structured paper," and "structured foil." 2. Structured webs are produced by (1) lamination, (2) coating, (3) coextrusion or (4) any combination of these procedures. |
| **Composite web:** A web that is not homogeneous in cross-section and is composed of two or more fundamentally dissimilar materials. | 1. Examples of composite webs: paper/foil, film/paper, film/foil, film/paper/-film/foil/film. 2. Composite webs are produced by lamination and/or coating. 3. Composite webs will not be referred to as "structures" or "composite structures." |
| **Lamination:** A web produced by combining two or more materials with an adhesive agent or by direct thermal bonding; also, the process of achieving such a combination web. | 1. Adhesive agents (e.g., starch-, wax- and plastics-based substances) can be deposited by extrusion, emulsion or solvent dispersion, or dry-resin deposition. Adhesive agents can be applied as heat-seal coatings prior to actual lamination. 2. One or both laminated webs can be a basic film or structured or composite web. |
| **Coating:** A thin layer of a wax and/or resin formulation deposited on a substrate; also, the process of adding such a layer to a substrate. | 1. Coatings are applied by (1) emulsion dispersion, solvent solution or powders, (2) extrusion or (3) waxing. |
| **Coextrusion:** A combination of two or more thermoplastics extruded as an entity by special dies or made by combining extruded thermoplastics before they harden into films; also, the process of combining two or more molten thermoplastics extruded from film dies. | 1. Coextrusions, in effect, are extruded structured films or heavy-duty films. An exception is a PP/PP/PP coextrusion intended to provide a "perfectly" neutral polypropylene film for subsequent biaxial orientation; this might be called a basic-film coextrusion. Typical coextrusions include two-layered films (LDPE/HDPE), three-layered films (PE/saran/PE or LDPE/MDPE/EVA) and—experimentally—four- and five-layered films. Still more complex coextrusions are possible. |

From *Modern Packaging*, Oct., Nov., and Dec., 1967. With permission.

proprietary to supplier and user. Where important and appropriate, some conclusions as to the nature of the packaging are drawn based on professional observation.

Because machinery and materials are so closely intertwined, the equipment used, usually used or probably used is described. It would indeed be a happy circumstance to be able to wrap up flexible packaging of specific food product groupings into neat little paragraphs. In each section, a summary of the significant materials, problems and trends is included, with the trends couched in the comfortable vantage point of today.

# BAKERY PRODUCTS

## Summary

Vectors of quality deterioration in baked goods are primarily associated with staling which is a series of inherent biochemical reactions. Packaging can have little beneficial effect on retarding this major cause of initial quality loss. Because staling is so rapid, bakeries are located close to the point of consumption and produce on a daily basis. Thus, protection requirements for baked goods are minimal; because of volume, packaging material costs are under great pressure to decrease.

Convenience packaging requirements have dictated the widespread use of tie-top reclosable polyethylene bags for bread and roll packaging. Although costing more than a polyolefin overwrap, polyethylene bags are relatively low cost when compared against other material possibilities.

Sweet goods, which have higher fat and sugar contents, require more greaseproof and release properties in packaging. Individual portion packaging has increased, leading to several different types of packaging: cellophane overwraps, heat sealed cellophane packages, die-folded aluminum foil wraps, and, now, thermoformed cups with heat sealed tops.

Equipment for current bakery packaging is generally based on machinery developed in other industries, and so has not always been well-integrated into bakery operations.

Published packaging developments in the bakery field do not forecast any significant innovations for flexible packaging in the future, but this observation must be couched with the knowledge that there is much unpublished activity.

## Introduction

Soft baked goods, such as bread, rolls, cakes and pastries, contain hydrated starch which revert to the crystalline state in staling. Since a substantial portion of the desirable properties of baked goods is in the swelled starch, alteration into a less tender form represents an important quality loss. Staling reactions proceed rapidly within hours, for changes perceptible to the senses take place.

The staling phenomenon is an exception to the rule that the rate of reaction decreases with reduced temperature; staling increases with decreasing temperature and sharply increases across the freezing point transition. Thus, if only staling were involved in bread quality retention, the optimum storage temperature would be above ambient. However, at elevated temperatures, loss of moisture and desirable volatile aromatics increases, and so room temperature conditions appear to be optimum for bread.

In sweet goods, the fat content is higher, retarding the staling process, which still occurs within hours. Reduction in temperature, however, helps retard changes, but unless below-freezing temperatures are employed, only a few days of salable shelf life are possible.

Because of the short duration of soft bakery goods salable quality retention, an industry system based on local baking and daily delivery and shelf stocking has evolved. Bread, rolls, cakes, doughnuts and pastries are prepared for delivery during the same 24 hours to the retail outlet, the HRI (hotel, restaurant, institutional) establishment, or directly to the consumer. Thus, the concept of "freshness" is imparted to the product, its baking and its distribution.

Many local and regional bakeries supply the daily consumer demands; some of these are part of large chains of similar bakeries located throughout the country and operating on an identical basis using similar raw materials and supplies. Some of the chains are composed of independent bakeries, each supplying the specific region, but they take advantage of the combined purchasing power of a large multi-bakery operation. In addition there are numerous small independent bakeries which package individual products to consumer order. More recently, the concept of "freshness" has led to increases in bakeries located in or near supermarkets to bake to order, or to prepare on an every-few-hours basis.

The most important baked product is bread. Bread and other yeast-raised non-sweet baked goods are the most susceptible to staling, and so are prepared and delivered on an every day basis.

## Packaging Requirements

Because bread's salable shelf-life is limited by staling, packaging can have relatively little positive effect on quality retention. On the other hand, packaging that encourages moisture loss is to be discouraged since water loss contributes to quality loss. Hermetic sealing to assure against moisture loss would not benefit bread during its extremely short shelf life. On the other hand, consumers often freeze packaged bread, and so some moisture protection to reduce freezer desiccation is desirable.

Bread does not require a highly greaseproof package. Bread's low fat content and short distribution life also mitigate against the need to protect against oxidation. Some marketers believe the visibility of bread and rolls is highly desirable.

Although the aroma of baking bread is traditionally a prized experience, commercial bakers have not been able to capture this desirable characteristic. Loss of aroma is not to be encouraged, but neither must any special properties be built into packages to assure against loss of aroma. In contrast, the flavor of commercial bread is so delicate that it is subject to pick-up from external sources. Thus, flavor barriers must act to exclude undesirable off-flavors.

Because bread is essentially an every-day consumer purchase containing great quantities of air for texture, it has a high physical volume. Bread, therefore, requires large amounts of packaging material. Most bread, however, is a commodity subject to severe competitive pricing pressures. Since the retail price is almost fixed because of the lack of differentiation between brands, the cost of packaging materials is under intense downward pressure.

Bread is a soft but somewhat elastic food which can be distorted and damaged by externally induced stresses. Ideally, bread should be protected against these dynamic shocks and strains by rigid packaging. Since bread must be delivered daily and is generally handled by an employee of the bakery, protection against physical damage is built into the distribution system. Carts and vehicles with racks, and returnable multi-wall corrugated single-tier cases are commonly employed. Driver-salesmen responsibility includes assurance that shelf-stacking does not induce crushing. Loaves are light weight and can be stacked within reason without visible damage.

Building physical protection into secondary packaging and distribution allows the primary packaging to be the lowest cost possible.

The higher fat contents of cakes, doughnuts and pastries, coupled with the short shelf life, lead to a requirement for greaseproofness in packaging materials.

## Packaging

Simple uncoated paper would absorb moisture and tend to stick to the surface of the bread leading to difficulty of removal. Coated or hard finish papers have thus been used for many years for bread wrapping. Retail bread bags are made of dense paper.

Waxed paper formerly provided containment, protection against moisture loss, some stiffness for stand-up qualities of wrapper side walls, for machinability of materials on packaging equipment, and a white base for a printable surface. Cellophane was the next big bread packaging material. Although printable, cellophane's decorative properties are best displayed by backside printing. Since conventional printing inks are not Food and Drug Administration approved for direct contact with food, backside printing is not possible without a packaging medium between the printing and the food. Duplexing is expensive.

This graphics requirement evolved into printed glassine or paper bands or end seals. End labels helped to seal ends which were otherwise difficult to seal because bread's softness did not provide a solid backing for heat sealing mechanisms.

The search for lower cost overwrapping materials in the 1950's and 1960's centered around thermoplastic polyolefins. Polyethylene was cheap but hazy in appearance and difficult to machine on overwrap equipment. Polypropylene film was found to be structurally adequate in thin gages, as it was as clear or clearer than cellophane, and possessed somewhat better machinability than polyethylene. More important, polypropylene in the acceptable gages was cheaper than any other usable film. Attempts were made to coat it for better machinability and moisture protection, and to strengthen it so that thinner gages and, consequently, cheaper per loaf packaging could be obtained.

One consumer convenience factor was always under study: the ability to open the package, remove part of the contents and reclose it so that the product would continue to be protected from moisture loss, a significant deteriorative vector in the home. A variety of peelable and restickable end labels were tried. Experiments, however, centered around methods which would accomplish the objective without raising the cost.

The interest in reclosable bread packaging led to development of a system for placing bread in tie-top polyethylene bags, much as soft goods, toy assortments and candy bar ten-packs had been packaged for some time—a prime example of transfer of a concept from one area to another.

The unusual aspect of the concept was that the packaging cost went up with its use, in contrast to the prior downward trends. Consumers found the plastic-coated wire ties to their liking, however. The bags, although made on three sides with hot-wire methods and efficient in use of polyethylene, required a rather long tail for the bunching and for wicketing on the packaging equipment. Thus, a large area of flexible film continued to be used for a loaf of bread.

Polyethylene could not be printed with fine detail without significant upcharge. Because of the film haze, however, topside printing was even more satisfactory than backside. Polyethylene perhaps was used in the original bread concept because this was a direct transfer from other areas of packaging. Polyethylene has, however, been retained because of its combination of cost, machinability and protection.

Rolls are converting into tie-top reclosable soft film bags. Even when the reclosable feature is not incorporated, soft film perimeter-sealed overwrapping is employed. The polyethylene bag has had such universal acceptance for

bread that it is now being used for all bread wrapping with confidence in consumer acceptance.

Most of the bread is packaged in bags fabricated from 1¼ mil low density clear polyethylene, with some being made from 1½ mil clear polyethylene. (As in many disciplines, designations vary among industry sectors; polyethylene film gage is most often designated by one mil equal to 0.001″).

Most closures now are plastic or paper coated metal ties. Some premium breads are first waxed paper overwrapped and subsequently bagged.

English muffins may be packaged directly into polyethylene bags or, as is more common, placed in paperboard trays which are inserted in the reclosable polyethylene bags. The paperboard serves as some structural protection for the relatively short textured English muffins.

Sweet goods packaging is far more variable than bread packaging because of the increased requirements and because of the greater variety of products manufactured. Cakes, for example, are definitely more susceptible to crushing than bread and, as a result, are most often packaged in rigid paperboard cartons which may contain a transparent cellophane window or are overwrapped in a protective film, such as heat sealing MS cellophane. Graphics on the overwrapped carton may be either on the carton itself or on the backside of the film.

Pies are often baked in aluminum plates which act as the package base after baking. Top protection is afforded by a paperboard or windowed paperboard carton, or, on occasion, by a cling film cover such as saran, or a transparent thermoformed polystyrene cover which the aluminum foil base fits.

Although portion packaged sweet goods have been on the market for several decades, only recently have they become a growth area. This phenomenon stemmed from the bold marketing move to multipack them, a concept transfer from beer, soft drinks and candy. Once multipacking had created a market awareness, the demand for individual units increased and was boosted by more effective point-of-purchase merchandising.

Iced cup cakes contain considerable fat in both frosting and crumb. Icing release from packaging material can be a serious problem;

polypropylene film or PDVC coatings are often used for release purposes. Furthermore, chocolate-flavored cup cakes can easily contaminate other baked goods with unwanted chocolate flavor. Cup cakes are easily crushed. In addition, size and configuration vary from unit to unit.

Packaging is on a one-side grease resistant coated die-cut paperboard sheet. Cup cake pairs are placed on the board and overwrapped, using MS cellophane in a die fold configuration which allows the heat sealing to occur at the base. The board, being larger than the dimensions of two cup cakes, allows for stretching of the cellophane over the top edge of the icing and thus minimizes side-to-side crushing of the cup cakes. The film is transparent to allow product visibility, moistureproof to minimize moisture loss, although the fold-seal is not a total closure, and resistant to passage of fat.

Other filled portion-pack sweet good pairs are packaged almost identically with the MS cellophane, also serving to retard sticking of the sugar frosting to the film. Polymer coated cellophane might have better release properties at a price premium.

Some bakers employ topside overall printed cellophane without the base board, obtaining a tack seal on the bottom through heat application. A number of single unit sweet baked goods are pouch-style packaged in cellophane. While effective seals are a desirable end result, they are probably secondary to the efficiency of continuous motion equipment used.

Some of the individually wrapped baked goods have overcome the problems of closure and sealing by taking advantage of the dead fold characteristics of aluminum foil. Since only minimum protection is desired from the foil, it serves as an interesting graphics background and as a good greaseproof material. Numerous individual iced cakes are die fold wrapped in surface printed, 0.00035″ aluminum foil. The bottom die fold is a series of overlapping crease folds which retain their position because of the dead fold characteristics of the surface aluminum foil.

The high fat content of pastry on individual fruit pies has dictated that a more fat resistant wrapping be employed. A frequent package is a surface printed preformed bag of transparent glassine. Alternatively, pouches are formed on the packaging machine from flat roll glassine.

Because of the protection afforded the baked product by its primary package, the need for added protection in multipacking is negligible. Thus, most multipacks for individually wrapped portion-packaged sweet goods are in printed paperboard cartons, although a few are in paperboard cartons with backside printed film double point end fold overwraps.

A recent means for packaging these products is to nest them in individual cups of thermoformed plastic. At present, a commercial test involves the use of preformed two compartment nested oriented polystyrene containers. A peelable transparent soft film is heat sealed to the flat lip to seal the cakes in the package. There is sufficient rigidity to allow stacking, and the lips, normally wasteful of heavy gage rigid thermoforming stock and difficult during multipacking, serve to separate the packages from each other. Sealed rigid polystyrene, while not an ideal moisture barrier, offers more protection than conventional cellophane overwrapping.

There have been several reports on the use of French Rilsan Nylon-11 which allows vacuum packaging and subsequent infrared heat sterilization of cakes. The process is claimed to allow a shelf life of several months to the cakes.

ICI (England) has had some success with baking in paperboard cartons lined with their TPX methylpentene heat stable polymer.[50]

**Equipment**

Systems for total packaging of bread and other baked goods are not well-developed. Individual machines for packaging are in various stages of sophistication, but few are well coupled with the needs of bakery production or converter operations.

The longest term development was for bread overwrapping equipment such as the Oliver Machine Co. machinery (which, incidentally, is also available for one and two slice bread wrapping). Such machinery had the advantage of being used in many other industries and thus being refined by the many interests concerned with its efficiency.

Equipment used for polyethylene bagging prior to the bread explosion, however, was inelegant, but perfectly satisfactory for the industries it served. Generally they had low labor costs. Most was (and still is) semi-automatic, requiring liberal use of manual labor. The machinery employed preformed bags stacked on wickets. Bags were opened by an air blast, filled manually and torn from the wickets by either the filling or by some mechanical or manual action. Tying was automatic or manual. Speeds were paced by the ability of the operator to fill the bags.

This type of equipment served as the basis for the AMF Mark 50 which became the standard of the bread packaging industry. Essentially, the inexpensive (under $1000) Tele-Sonic type equipment, used so widely for small parts, was mechanized and automated into the bread bagging machine. Many differences, of course, exist between the two machines, but the basis of using preformed bags on a wicket, blown open, filled and then torn from the wicket is the same in the two units.

A preformed bag machine offers versatility in that many different sizes of bag and, consequently, product may be used. More recently, others, such as the Pneumatic Scale Corp., have introduced their own version of the equipment, this one focusing on bun and roll bagging.

FMC Corporation, recognizing the inefficiencies of intermittent motion equipment, and of the user having to purchase premade bags, designed a unit which forms and fills bags from less costly printed roll stock. Because the wicket operation is eliminated, less material is used for each bread wrap. Further, this type of equipment should run at higher speeds, but should be more expensive than the premade bag machine because it includes the bag making operation.

Although machinery companies are generally small with limited design and engineering resources, the magnitude of the bakery packaging industry leads to the conclusion that the ultimate in bread packaging equipment, even for reclosable bags, has not been attained.

The production volume of full-size cakes is too low to have warranted automatic filling equipment, although overwrapping is performed on automatic equipment. Carton erection may also be accomplished on automatic lock corner equipment, such as Kliklok, Peters, U.S. Automatic Box or Sprinter.

Cup cake portion packaging has been performed for many years on Battle Creek equipment designed for gentle handling of the product and equally careful forming of the film around the product. This equipment employs pusher mechanisms which restrict it to stiff films such as cellophane.

Individual pouch-style cake wrapping uses the venerable Hudson Sharp Campbell wrappers which pull roll stock through and over food units riding under on conveyer flights. The film continuously wraps around the products and is sealed in a continuous line. Rotating sealer bars simultaneously form two end seals and cut them apart. The continuous motion allows for significant increase in speed (up to 200-300 pm) over intermittent motion machines.

Recently, similar type equipment has been introduced by SIG (Switzerland), a very high speed unit; Package Machinery Corp. (Forgrove Rose from England); and Doughboy Industries (Mustang), a far less expensive unit.

Die wrapping has become the province of Wrap King, now part of the Crompton and Knowles organization. The nature of the fold limits the speed of this equipment.

The machines enumerated are only typical of equipment used in the bakery industry. The list is by no means complete, nor does inclusion in this list imply superiority over any not included. The listing is to describe the types of units on which flexible packaging material is used, and, as such, is an integral part of flexible packaging of foods.

## BISCUITS AND CRACKERS

**Summary**

Cookies and crackers are longer shelf-life products which are under good distributional control by the bakers and packagers. The products are dry and have high fat contents. Crackers lose crispness when they absorb moisture and, therefore, should have protection against environmental water vapor. Thus, packaging for these products includes a fat barrier and some moisture protection.

The products are generally fragile and require structural protection. Cookie and cracker packaging, therefore, is a combination of flexible film and paperboard. Glassine and waxed glassine are among the more commonly used materials.

Thermoformed plastic trays are replacing glassine cups for compartmentalizing cookies.

These trays are film overwrapped for moisture protection. Similar overwrapping is used for corrugated glassine lined paperboard trays.

Crackers and cookies which can withstand physical abuse are dumpfilled into glassine lined paperboard cartons made on double package makers. Waxed glassine, folded over for closure, provides sufficient moisture and fat protection.

Cookies, especially those with precise shapes and sizes, are being stacked into lined standup bags and into film overwraps which are placed in paperboard cartons. The latter slug wrapping provides protection and reduces labor costs.

Some rotary-die formed cookies and soft marshmallow cookies are packed in header-type cellophane bags after stacking in partitioned compartments made from glassine or paperboard.

Printed paperboard cartons with protective inner flexible packaging made on automatic equipment is the method which appears to be the leader today. Some sealed thermoformed plastic trays will doubtless be used in the future, but not to any great extent because the products do not require great protection.

If the present distribution system is broken, there will be different shelf-life and structural requirements, and some changes in packaging may be expected. The investment in equipment, however, makes rapid changeover unlikely.

**Introduction**

The low moisture content of cookies and crackers leads to extended shelf life for these products. Cookies have the additional benefit of sugar to retard microbiological deterioration, particularly in the higher water content soft cookies, such as marshmallow. The fat which imparts shortness to crackers and cookies, however, is subject to rancidity which can be accelerated by increased moisture. Duration of retention of initial quality for these products is measured in weeks and months.

Since cookies and crackers, for marketing purposes, have long shelf lives, bakers manufacture at central bakeries and distribute over long distances. The ability to manufacture at centralized sites has led to the concentration of manufacturers in this industry. Three or four

large biscuit and cracker bakers are able to supply almost the entire country from their multiplant operations.

In addition to the major bakeries, there are a large number of smaller operations which make specialties or which cover a limited geographic region. Thus, almost all retail marketing areas are served by a few national manufacturers, each with an umbrella brand, plus several regional, local and private label bakers.

Multiplant bakeries employ similar distribution systems which are between those of bread bakeries and grocery manufacturers. Although many attempts have been made by the biscuit and cracker bakers to eliminate driver/sales/service/men, no one has been successful. Cookies and crackers are delivered to the retail outlets and placed on the shelves by driver/salesman. Some specialty cookies also travel through conventional dry grocery distribution channels.

While the present common distribution procedure is costly to the bakeries from a labor standpoint, it has advantages not enjoyed by other food manufacturers. Naturally, the shelves are stocked with both the mix and quantity of products the baker believes will move best in a given retail outlet. Stock rotation, so often a problem with other foods, is a regular duty of the driver-salesman. This control, exerted by bakers over their products, allows for extensive pilot and market testing of both products and packaging.

Since processing is common to baking of both retail and HRI products, the national biscuit and cracker bakers package many of their products for the hotel, restaurant, and institutional businesses. Many of these products are distributed by the same vehicles for the retail business, but rather than servicing individual restaurants, wholesale suppliers receive the goods.

Products of biscuit and cracker bakers may include snacks, cereals, candies, and short shelf-life bakery products. Many bread bakeries complete their sweet good lines with limited lines of specialty cookies, most of which are of the soft variety with relatively short shelf lives.

The basic product lines of the national biscuit and cracker bakers consist of saltine crackers, cocktail crackers, vanilla wafers, cookie sandwiches and both die-cut and wire-cut cookies. Total product lines may run into the hundreds, with but a dozen or so having been in the line for many years. The remainder of the line is a constantly changing group of cookies, crackers and snack crackers varied in shape, flavor, texture and inclusions. With controlled shelf distribution, new product ideas can receive rapid market distribution and test results. Thus, the solid base of saltines and die-cut cookie sandwiches is backed up by a broad range of flavored crackers, sugar wafers, cookies of various flavors with and without icing and nuts, graham crackers, marshmallow cookies, and sandwiches. More recent additions to product lines have included sesame seed crackers, rye crackers, pizza flavored snacks and peanut cream sandwiches.

Added to these baked products are foods which may also undergo processing similar to baking but which are not usually designated as cookies and crackers. Long shelf-life chip-like snacks, enrobed sugar wafer sandwiches, tart shells, and prepared pie crusts would be included in this category.

Those products which are conventionally classified as cookies and crackers are included in this discussion. Product lines which normally fall into other categories are discussed in their respective sections.

**Packaging Requirements**

Because cookies and crackers are low water content products and usually deteriorate by the incorporation of much added moisture, exclusion of water vapor is a packaging consideration.

Major consideration is given to the visual appearance of the package. Crackers and cookies are generally high fat products subject to oiling off with temperature fluctuation. Many cookies are enrobed with high fat coatings or have high fat cream centers. Flow of liquid fat from the product under these common circumstances would lead to spreading into packaging materials which are not grease resistant. Fat staining is unsightly, makes the package appear old, and by increasing exposure of the fat to air can lead to rancidity. Thus, all primary packaging has an initial requirement of grease resistance.

Crackers are susceptible to moisture pickup and subsequent loss of crispness. Thus, crackers are protected against ingress of environmental moisture. Because of distributional control, however, hermetic sealing is used only infrequently. Further, many studies have demonstrated the desirability of breathing for packages containing cereal products. The flavor of saltine and unsalted crackers is mild and alterable by intrusion of unwanted flavors. Greaseproof packaging, fortunately, generally protects against off-flavors from external sources.

By the same token, the flavors of some cookies can contaminate other products. The same greaseproof packaging minimizes loss of desirable flavors, as well as transfer of flavors to other nearby products.

A few cookies, such as marshmallows and jellies, have relatively high water content which, if lost, leads to sugar crystallization and hardening of the soft centers and toppings.

Although fat oxidation would be an ultimate problem with these long shelf-life products, the problem is usually so infrequent because of distribution control that exclusion of oxygen is not warranted. This stability is further enhanced through the employment of hydrogenated fats and antioxidants as ingredients.

Crackers are fragile and so warrant structural protection to minimize damage. Because of light weight and small units, however, structural rigidity is usually not found. Often, the crackers are small so that they can be dump filled and handled as bulk products.

Wire-cut cookies are made from doughs which allow natural spreading during baking. This can lead to thin, irregular edges and odd-shaped products susceptible to fracture and crumbling. Protection against crushing and shock damage is thus more important with wire-cut cookies than with dump filled snack crackers. However, the packager is probably limited in providing such protection for the product by its irregular shape. Again, the problem is not of such importance that great care is always taken to provide this structural rigidity.

Die-cut cookies, on the other hand, have precisely defined sizes, shapes and patterns. The pattern is often the trade mark for the product, and often contains the manufacturer's brand name. Because die-cut cookies and cookie sandwiches are the most attractive and expensive products in a baker's line, care is exercised to assure against crushing, chipping and breaking these products. Dividers, compartments and other such structural devices are costly, but have the advantage that they occupy space which would otherwise be taken by more costly product.

Soft goods, such as enrobed marshmallow cookies, demand structural protection because they are so fragile. Appearance of these products is a desirable marketing tool which is maintained throughout the marketing life.

Packaging costs are significant. Retail pricing is not so rigid that there is no flexibility. Furthermore, there are many exclusive products in which packaging costs are less important than they would be were they commodity products. As will be detailed below, a large number of products are designed to employ common packaging materials and machines, thus effecting packaging cost savings in the most important areas—the packaging operations themselves.

### Packaging—Crackers

Crackers are credited with having been the first food products to be consumer packaged. The packaging which allowed removal of crackers from the barrel was the basis for most food packaging.

Extensive testing has demonstrated clearly that separation of food from its environment is best accomplished by placing the barrier next to the product. Formerly, the impermeable packaging materials were on the exterior where it was mechanically simple to place them. When experimental results indicated the benefit of positioning the barrier adjacent to the product, machinery was designed to achieve this objective. Because of the limited number of potential customers for such equipment, machinery manufacturers did not hasten to develop appropriate equipment. The biscuit and cracker bakers, in many instances, found that it was advantageous to design and build their own equipment.

Early cracker packaging involved combinations of waxed papers and paperboard cartons. Waxed papers could be made from conventional MG sheets, but wax and fat could be mutually soluble, and so fat could migrate through the wax which served as a moisture barrier. Petroleum-based waxes were fine mois-

ture barriers on the surfaces of flat paper sheets. Waxes, however, were brittle and broke at folds and creases. Cellulose fibers from the paper protruding through the wax acted as moisture wicks to transfer water through to the crackers.

These deficiencies led to the employment of glassine as the waxing substrate, a practice which is continued today. Glassine is a hydrated short fiber length paper with dense formation that makes the sheet inherently greaseproof. Supercalendering after papermaking gives glassine a highly polished surface which accepts coatings such as wax. Thus, although glassine has no moisture barrier properties of its own, it is an excellent medium to carry the moistureproof properties of wax. Glassine complements its substrate status by having its own greaseproof properties and by being very machinable.

Heat is used for sealing packages fabricated from waxed glassine. Melting of wax on equipment leads to a build-up and the need for regular operator maintenance. This is a problem area.

Waxed glassine, although a distinct improvement over waxed MG sheet, had some deficiencies. The commercialization of polyethylene and other rubber-like polymers led to the incorporation of additives into the waxes to reduce crease cracking and improve overall package performance. The introduction of tacky, flexible microcrystalline wax allowed wax to be used as an effective laminating agent. Several sheets of waxed glassine could be married with limited quantities of microcrystalline wax to build up the amount of wax which could be incorporated into the sheet to reduce moisture permeability. Simply adding more wax to the surface of glassine has led to exponential increases in wax build-up on packaging equipment.

In the early 1960's, ethylene/vinyl acetate copolymers were introduced into waxes to improve their flexibility, to the point where they could be handled much like plastic films. These additives also allowed for alteration of wax melting points and improved the surface glass of the sheets.

The choice of amber, bleached or opaque glassine is a marketing decision. Amber is the least expensive, opaque the most expensive.

Bleached and opaque grades have a cleaner, more sanitary appearance. All three are used by various cracker bakers for packaging, usually in single ply two-side overwaxed form, using 25-30 pound base stock. This is about the least expensive moisture barrier sheet which can be used on existing cracker packaging equipment.

While the barrier sheet has undergone several changes, the most profound changes have been in the package form. Originally, waxed sheets were employed as overwraps on chipboard cartons often with folded waxed glassine liners for greaseproofness. Once a cracker package was opened, it could not be effectively reclosed. Cracker packagers broke the traditional one-pound packages down into four quarter-pound packs, each overwrapped with waxed glassine. Some used long stacks of individual crackers with face-to-face long and end seals all folded over; some tiered a small quantity of four-cracker units and overwrapped them using a double-point end fold. Now the consumer could open a small quantity at a time and use the contents without the remainder being subject to moisture problems.

Sealing the package to the contents led to greatly improved moisture protection, and the need for a waxed paper overwrap was obviated. Printed one-side bleached or coated cylinder board cartons are used in place of overwrapped chipboard. Easy opening and reclosure features are incorporated into the carton. Some waxed paper overwrapped chipboard cartons are still used commercially.

Borrowing techniques from the bread packaging industry, some bakers pack slugs in preformed tie-top reclosable polyethylene bags. This bagging adds to convenience and protection and should allow for reduction in wax weight on the glassine.

The preformed bag was soon confronted with competition from the bag formed in-line from roll stock.

Waxed glassine met some competition from cellophane which had the advantage of transparency. Cellophane has been used as a replacement for preformed waxed glassine bags, but not on a bag-for-bag basis. Preformed cellophane bags would be more expensive than similar glassine bags, but cellophane wrapping on newly designed equipment would be less expensive than preformed glassine bags and semi-automatic or manual filling. The replacement has been made in some of the few operations which had persisted in the use of preformed glassine bags.

Other plastic films have been tried to replace glassine, but less expensive films are not readily machinable on existing equipment. Heavy investment in cracker wrapping equipment capable of handling plastic film could not always be justified because of the dynamic state of saltine packaging, and because saltines are the commodity side of the business. Besides, some of the cracker packagers have substantial engineering investments in equipment of their own design.

Biaxially oriented polypropylene is machinable on newer European equipment, is less expensive than glassine in the gages required, and has the desired moisture barrier properties in the film. Coating adds to the attributes. Polypropylene is also a greaseproof material, and is thus being tried for slug wrapping of crackers.

Individual HRI packaging of saltine crackers has been, and continues to be, top side printed MS or polymer-coated cellophane in pouch style packages. Transparency is important for identification; printability is necessary for use in hotel and restaurant operations, and MS and polymer-coated grades offer both moisture and grease protection as well as heat sealability. Cellophane is machinable especially for smaller, comparatively intricate packages. Unmounted polypropylene sealing is not easy and, even when technically effective, may not be sufficiently attractive for display purposes. Polyethylene is too soft for easy machining in high speed operations required for cracker packaging and lacks transparency.

**Packaging—Snack Crackers**

While saltines and cocktail crackers have the greatest volume of biscuit and cracker packaging, snack crackers, or their equivalent, have the greatest product variety. Most of these are small and flavored, with less shortness and, hence, less fragility. Because of size and strength, they need not be carefully stacked to minimize physical damage. Snack crackers are usually dump filled. Due to the difficulty of shaking down crackers during filling and closing, packages are filled to volume capacity to shake down during distribution and leave a headspace. This shaking and headspace allow for considerable movement and abrading, and potential for damage and breakage. However, snack crackers are sufficiently durable to withstand this type of treatment. The built-in resistance to fracturing is an aspect of texture which apparently lends itself to desirable sensory characteristics.

Retention of these masticatory properties depends on exclusion of moisture which could be, and occasionally is, achieved by hermetic sealing in a highly moisture impermeable film or lamination. Again, baker control has allowed the most common packaging to be glassine/wax lamination liner in a paperboard carton.

Glassine, in proximity with the product, retards fat penetration, retains aromatics in the crackers and excludes potential off-flavors which might arise from the paperboard. Use of glassine as an odor barrier allows the employment of newsback cylinder board with its occasional problems of off-odors arising from the furnish.

The glassine liner may be formed using adhesive on an overlapping long seal, and so the glassine is one-side waxed. The bottom seal may be a foldover, or an adhesive seal. After filling, the top closure is generally a multiple foldover and not a seal. Since there is no backing, a seal would be almost impossible to effect. Tests have indicated that the multiple foldover creates a tortuous path for moisture and so effectively protects the contents.

When added moisture protection is required, a waxed glassine or glassine lamination may be face-to-face sealed and folded over. Opening may be by simple peeling of the wax-to-wax bond.

Some bakers have gone to great lengths to provide moisture protection for snack crackers by using paper-aluminum foil combinations and sealing them in the carton. Some have gone to even greater lengths by combining paper, aluminum foil and polyvinylidene chloride in a single lamination to create a sealed bag. Aluminum foil and PVDC are two effective moisture barriers. The brittleness of these two barriers often limits their use on vertical form, fill and seal equipment. Forming collars tend to stress crack the materials being formed, and thus break the integrity of the coating. With the suc-

cess of glassine carton liners, however, in protecting against moisture, there have to be some very unusual protection requirements or some very strong marketing reasons to employ such an expensive internal bag.

Some snack bakers have gone to even greater extremes by using preformed bag-in-box combinations utilizing paper/aluminum foil combinations. The products contained, however, are more often snacks than crackers, and have a far higher fat content. Bags for bag-in-box systems are preformed face-to-face side seal paper/aluminum foil laminations. Bags in flat form are then adhered to the interior of knocked-down paperboard cartons. When the carton is snapped open for filling, the bag opens.

Cartons are used for structural support. This allows stacking without compressive top-to-bottom or side-to-side damage. Although waxed glassine could be used for portion packaging, marketing considerations have dictated transparency. Surface printed MS and polymer-coated cellophane bags are most often used for small portion packages of snack crackers. These are generally sealed pillow packages which make most efficient use of material to obtain internal volume, although there are some problems with coating cracking and with obtaining complete seals at the crossover of end and back seals.

In recent years, a limited number of premium snack crackers have been packaged in stand-up bags. These bags appear to be preformed gusseted bags with printed litho stock exterior, paperboard liners for structural support, and sealed aluminum foil bag liners. Some of these are laminations of PVDC or polypropylene and paper with foldover closures at the top. These stand-up bags are necessarily more expensive than waxed glassine-lined paperboard cartons, but represent a means to gain a merchandising advantage over the paperboard carton systems.

Some snack crackers are oriented and stacked in waxed glassine lined paperboard cartons. The paper surrounds the crackers but is neither sealed nor fold-closed. Because the glassine in these packages is open, it has low moisture impermeability. Thus, the carton is overwrapped with printed waxed paper to provide moisture protection.

Sandwich crackers containing cheese, peanut

butter and similar fillings have traditionally been in portion packages for the vending trade. Recently, these portion packs have been multi-packed. Generally, the immediate package is MS or polymer coated cellophane wrapped around one or more sandwiches. The wrapping is either double point end fold overwrap or pouch-style.

Cellophane acts as a moisture, odor, and grease barrier. Generally, if MS grades of cellophane are used, the gage is 195 (i.e., 19,500 sq in. per pound unprinted). If polymer coated grades are used, the gage is usually thinner, 210 or 250 (i.e., 21,000 or 25,000 sq in. per pound). Polymer coated cellophane offers greater moisture protection than MS (moisture-proof, heat sealing) grades. Both are apparently satisfactory with the economics being approximately equivalent taking into account the gage differences.

### Packaging—Cookies

Wire-cut cookies are packaged in much the same types of materials and forms as are snack crackers. Vanilla wafers are typical of wire-cut cookies dump filled in waxed glassine lamination lined paperboard cartons. As with crackers, waxed glassine acts as a moisture barrier. Glassine is also a fat barrier to retard staining on the exterior paperboard carton.

Cookies are larger than snack crackers and so can be more fragile. Thus, relatively few wire-cut cookies are dump filled. Many are carefully packaged in stacks to reduce possible damage. Some of these cookies are oriented into packages which contain corrugated uncoated glassine liners. The glassine serves as cushioning to minimize crushing, cracking and chipping.

One of the major cookie bakers employs stand-up bags for wire-cut cookies. The cookies are oriented and stacked in gusseted, printed, paper bags with polypropylene liner affixed to the interior. The polypropylene acts as a moisture and fat barrier. Stacking of the cookies forms a fairly solid block for stability and reduces physical damage. Bags are preformed and manually filled to assure proper stacking.

Rotary die-cut cookies and cookie sandwiches are packaged in a relatively wide variety of packaging types. Probably the most rapidly growing of these is thermoformed rigid poly-styrene trays which are subsequently over-wrapped. Formerly, rotary die-cut cookies and sandwiches were placed in fluted glassine cups which were in overwrapped paperboard cartons. Because packaging involved many pieces of material and manual insertion, the use of a single multicup insert was initiated.

Residual monomer in early polystyrene resins resulted in highly undesirable off-flavors in the contained food products. Cookie bakers did not generally attempt to use rigid polystyrene in these early periods, so most did not directly experience the problem. Resin manufacturers have since cleaned up their plastic.

Oriented polystyrene (OPS) in 7-20 mil (0.007" - 0.020") gages is thermoformed into multicavity trays. Each cavity is shaped and sized to hold the rotary die-cut cookies. The cookies are placed flat or on edge. In the latter case the trays cover only the lower portion of the generally circular-shaped cookies. Cookies oriented flat are stacked only to the level of the upper rim of the tray.

Edge-stacked cookies in OPS trays are most often in pouch-style overwraps using fin end seals and overlap or folded over bottom fin seal. Because rigid OPS is not a good moisture barrier, MS or polymer coated cellophane is the overwrap material. The film is either surface printed or has a paper label adhesive affixed to the top.

These packages generally use heavier gages of OPS since the trays impart all the structural rigidity and physical protection. Other materials, such as rigid polyvinyl chloride, would offer greater moisture barrier per unit gage, but thin out more than OPS in thermoforming. PVC's clarity is less than that of OPS. PVC is more expensive.

Although thermoformed plastic trays are enjoying considerable growth, most edge-stacked cookies today are still packaged in corrugated glassine lined paperboard trays. The paperboard may have the corrugated glassine laminated to the tray interior; such trays are formed, using glued or thermally affixed paper corner stays. The more frequently used tray is a separate paperboard tray with lock or glued corners and with die-cut single face corrugated glassine inserted into the tray. The glassine is almost always opaque white, brown or red. Filled trays are overwrapped with cellophane.

Since there is no lip, as in a thermoformed plastic container, a pouch-style wrap is not necessary.

The overwrapped tray is an inexpensive packaging system used for two extremes of cookie products. Less expensive cookies of large multiplant bakers are often packaged using the overwrapped tray system. Premium cookies of smaller bakers are also packaged using this system, probably because the equipment is relatively inexpensive and appropriate for smaller-sized companies.

The more common packaging system for rotary die-cut cookies is placement in compartments in hinged-top paperboard cartons. Thermoformed OPS trays are growing in importance as inserts for such cartons, but fluted glassine cups and paperboard dividers remain the principal system used at this moment. Because the paperboard carton provides structural support, OPS gages are relatively thin. Cartons containing vacuum formed plastic trays build the moisture barrier into an overwrap. Waxed paper and even aluminum foil / paper / wax / strike-through tissue are used as overwrapping materials. The flow of wax in thermal sealing tends to seal the closure fairly well. The aluminum foil lamination is more expensive than the waxed paper but has the advantage of imparting added graphic attractiveness and holding more wax closer to the product than waxed paper.

Because cookies do not require low moisture permeability, waxed papers are used more commonly as overwraps, when overwraps are used. In another method, the OPS tray is set into a waxed glassine sheet folded over the top.

Overwrapping is not commonly employed by the biscuit and cracker bakers for cookie packaging. Trays and overwrapped slugs and stacks of cookies are placed in lithographed cylinder board paperboard cartons, end-opening sleeves or hinge-top opening trays. The trays are formed by lock or glued corner methods, often on automatic equipment. Internal plastic trays are inserted and filled with cookies. The carton is then closed.

Slug or stack packaging of cookies is a growing method because of the availability of automatic collating equipment. Many rotary die-cut cookies are still manually removed from the production line and placed in preformed glassine bags which are manually folded at one end. The glassine, being uncoated, serves only as a fat barrier. Several of these glassine slugs are manually inserted in paperboard cartons.

In a more advanced method, human hands are used to transfer, count, and align stacks of cookies into flights which feed into pouch-style cellophane or polypropylene wrapping machines. Glassine, being short-fibered, may not be as machinable on this type of equipment as are the transparent films.

A more sophisticated European machine, which has been employed by English biscuit makers for some years, performs all the collating, alignment, removal and wrapping automatically and continuously. Again, cellophane or coated, biaxially oriented polypropylene is used. The relatively poor appearance of the polypropylene end and back seals is not visible on the inside of the carton.

Slugs made on this automatic wrapping equipment may be manually or mechanically placed in end or side loading paperboard sleeves.

Despite the increase in use of slug wrapping for rotary die-cut cookies, most such cookies are still not wrapped. The elegant appearance of fluted glassine cups to hold a limited number of stacked cookies or cookie sandwiches has been accepted for higher priced assortments. Other packs have simply not yet made the conversion.

Glassine fluting has an advantage over thermoformed OPS: it has more "give," and, as a result, occasional off-size of off-center cookies are not chipped or fractured by insertion into the cavity. The use of fluted glassine cups, glassine sheet liners and chocolate-colored paperboard dividers is declining because of the labor costs involved.

In some marketing areas, transparent bags with headers for cookie sandwiches are in demand. The cookie sandwiches are stacked in double face, corrugated, colored glassine U boards which are placed in bags. Since visibility of the rotary die-cut cookies is the attraction of this package, the bag is formed from transparent material. Because the film must serve as the total moisture, fat and physical barrier, a heavy material such as 140 gage cellophane, surface printed, is generally used. Cellophane is stiffer than true plastics and has the sparkle and transparency that marketing demands. Furthermore, cellophane can be printed by rotogravure methods while the plastic films are better printed by flexographic methods. In general, these methods give graphically inferior results (although flexography has improved in recent years). With the cellophane graphics competing directly against paperboard lithography, good reproduction is an important consideration. Closure is by heat sealed paper header which acts to seal the cellophane at the top and also adds a printed message to the top, or end, depending on stacking. The paper header also acts as a carrying device for the consumer to help prevent crushing or breakage and retain bag appearance. The internal corrugated glassine acts as cushioning and as a mechanism to hold two stacks in line and apart from each other.

Somewhat similar to this package is the system used for marshmallow cookies, most of which are packed in transparent bags. The same type of gusseted cellophane header bag is used, but the cookies are inserted into paperboard egg crate partition compartments. The paperboard may be chocolate colored to hide fat migration or coated to resist it. A corrugated glassine pad may be placed at the bottom to assist the bag insertion and to provide some cushioning. Again, heavy gage MS cellophane provides the barrier required.

A relatively recent category of enrobed marshmallow cookies, the giant-size units, are not packaged in compartments because of their size. Each unit is individually wrapped in a pouch-style package or placed in a preformed glassine bag, although the latter does not provide the needed moisture barrier. Minimization of moisture loss from the marshmallow cream is effected by the use of MS coatings. The packages are stacked into chipboard folding cartons overwrapped with waxed paper. Greaseproofness is also provided by the sealed internal packages.

Sugar wafers are always in sandwich form with cream centers. They are cut to regular sizes and shapes, and always have sharp, abrasive edges and sides. Sugar wafer sandwiches are easily stacked into compact tight blocks which can be placed in cartons or wrapped. One manufacturer employs a paperboard carton with an aluminum foil/paper/wax/strike-through tissue overwrap. The sugar wafer is a gross moisture absorber and a product which, when soggy, has poor texture. It is a commodity which generally has heavy price competition from private label bakers. Printed aluminum foil overwrap tends to impart brand identity to an undifferentiated food.

At the other extreme, sugar wafer sandwiches are blocked and wrapped in flexographically surface printed MS cellophane. In this more commonly seen package, the sugar wafer sandwich is a commodity and treated as such with little consideration given the protection or consumer convenience on reclosure. Inexpensive overwrapping is carried even further by using polyethylene as the flexible film.

## Equipment

For various reasons much cookie packaging equipment was designed and built by the engineering departments of the cookie bakers themselves. Perhaps the best known of these is the system for wrapping saltine crackers in waxed glassine. The fact that such machinery was not developed by those who specialize in this business is sufficient commentary on this situation.

Commercial stock equipment for cracker packaging often involves the combination of two units, one to collate the crackers and the other to wrap. Collation thus may be performed manually or semi-automatically if desired by the packager. Hayssen, Battle Creek, or Package Machinery Corp., make conventional stock overwrapping equipment which can be used for double point end fold wrapping; Hudson Sharp Campbell wrappers may be used for pouch-style packages. All require refrigeration units to chill the wax after tacking the seals.

Insertion of the slugs into polyethylene bags may be manual or may involve the use of an FMC Corporation unit which forms reclosable bags from roll polyethylene stock.

Portion packaging of two tiers of saltine crackers is usually accomplished on Hudson-Sharp Campbell wrappers which form pouch-style packages at upwards of 200-300 pm. Recently, SIG units, which operate at up to 480 packs per minute, have been installed for portion packaging. Portion packaging of cracker sandwiches is done on overwrapping equipment, such as Package Machinery Corp., units, which make double point end folds, or Lynch equipment which makes a bottom seal die fold. The Lynch units, which are common in the confectionery industry, place a base board under the cracker sandwiches from either die-cut or roll stock.

Dump filling of snack crackers into portion packages is performed on vertical form, fill and seal equipment such as FMC Corporation, Stokeswrap; Package Machinery Corp., Trans-wraps; Triangle; Hayssen; Rovema; and others from among approximately 80 makes on the market.

The packaging machine seen most often in cookie and cracker baking plant packaging operations is the Pneumatic Scale Corp. double package maker. This unit, which can be installed complete, is a large and somewhat complex piece of equipment, but one which has proven its reliability in over two decades of steady use. The double package maker (DPM) is difficult to change over from one size or shape to another and requires a large amount of floor space. DPM's are based on original European designs with the German Hesser being the only major alternative to the Pneumatic Scale equipment. Peters also has a packaging machine which inserts a glassine liner.

The DPM forms an internal paper or glassine carton liner over a mandrel and then forms the paperboard carton, discharging a lined paperboard carton for dump filling at the next station. The liner is formed from roll stock. The carton may be formed from flat blanks or from preformed sleeves.

After filling, the DPM closes the interior liner by sealing or folding over, or both, and then seals the carton top. These units can operate at 30-100 packages per minute.

Interstate Carton Corp. and Container Corporation of America both have bag-in-box systems which are only infrequently used by the cracker bakers.

Thermoformed cookie trays are usually wrapped on larger size Campbell wrappers which are somewhat fixed in package size. The speed of these continuous motion machines is more limited by availability of product than by equipment capacity. Speeds of 30-60 pm in these sizes are common.

Filled paperboard trays may be wrapped on double point end fold overwrapping equipment which has a speed capability of up to 60 pm.

Slug wrapping is a relatively new technology in the United States. Up to now, Campbell wrappers have been modified by bakery engineers to provide wrapping and sealing for the stacks. More recently, European equipment, which has been used for this purpose for a number of years, has been imported into this country. SIG equipment has been installed and apparently some tests have been run with Fore-grove Rose machinery from England.

Equipment for cookie bagging is generally lacking although this has been a growing field.

The DPM is evidently the only universally used packaging machine for cookies and crackers which has been refined for this product category. Most other packaging equipment has been adapted from other uses or industries. Only the SIG slug wrapper appears to have been developed for cookie packaging purposes.

This is another unfortunate example of machinery manufacturers failing to respond to the needs of the potential users. The engineering departments of most cookie and cracker bakers have limited resources. The product lines among competing bakers are quite similar, as is the packaging. Thus, some equipment should have been developed.

# CANNED FOODS

**Summary**

Apart from label stock, flexible films have the potential to be used in place of metal cans for some thermally processed products. They are already finding some use as inexpensive containers for aseptically processed fluids. Much experimental work has been performed on heat sterilization of food after filling and sealing in flexible packages. Ideally, such packages would provide better heat transfer and less thermal damage to the food. Film combinations available, however, are claimed to have seal reliability below the levels set by the program sponsor, the government. There is insufficient information published to critically review the flexican concept. Neither the government nor the few commercial interests has provided enough data, although there is no indication that they would not answer all questions put to them.

The use of shrink bundling of cans as a shipping case replacement is proceeding at a low rate. There are numerous problems associated with shrink wrapping of beer six-packs. The advent of the two-piece necked-in aluminum can has allowed some commercialization, but equipment problems are complex. Marketing pressures and significant increases in quantities of necked-in cans are required to advance beer can shrink six packing.

## Introduction

Three basic areas are discussed under this heading: multipacking of cans using flexible materials, aseptic packing of fluids, and flexible packaging of foods followed by heat sterilization. The last has been under study by the military for some years, and has been the subject of a limited number of publications. Little of a practical commercial nature has emerged from the extensive development activities. Neither has this effort resulted in process, product or package which has been acceptable to the government agencies who are sponsoring the work.

Aseptic processing and packaging of fluids has been commercial on a limited scale for some years. Most of the packaging for aseptic processing has been rigid, for example, cans, pressurized cans, glass bottles, and more recently, blow-molded plastic bottles. In the last two or three years, there have been some commercially successful developments of aseptic packaging in thermoformed cup fill and seal systems closed with flexible films, and all-flexible form, fill and seal systems. These developments are spreading into more uses and are discussed under the subject headings dealing with the products contained.

Multipacking of cans and bottles developed very rapidly during the 1960's. Formerly confined to a convenient means for the consumer to carry returnable bottles back and forth from home to store, the concept benefited from retailers' upwardly spiraling labor costs. Retailers could not afford to handle returnable glass in the volumes in which consumers were purchasing multipacks.

The non-returnable bottle brought with it a host of multipacking schemes. Simultaneously, canned beer and soft drinks increased dramatically, and with them, the number of methods

for their multipacking. Metal clips, compartmented paperboard, rigid plastic and flexible film were all tried commercially. The multipacking situation has developed into a relatively clear picture with flexible film being used for a small portion of the beer can part of the market. Most of the beer industry employs the Hi-Cone® rigid plastic carrier for its can six- and eight-packs. The smaller soft drink canning industry has generally confined itself to paperboard carriers. Beer and soft drink bottle multipack carriers are usually paperboard carriers selected from among the several systems available.

A limited number of juice, pudding and fruit canners have introduced multipacking of their products, usually of individual portion sizes in two's, four's and six's. Rather than invest in any new developments, the well-tried paperboard carriers and shrink film have been used.

There have been some commercial developments in the employment of shrink films in conjunction with corrugated trays in place of corrugated cases for shipping containers. At least one major canner has been making test shipments using this system for several years. These shipments are under the close scrutiny of governmental and transportation regulatory authorities. Competitors as well as corrugated interests have watched these experiments with concern, but there has been no significant visible expansion of the activity despite intensive efforts by a few of the shrink film manufacturers. More recently, several equipment makers have intensified their developments and marketing activities relative to these systems.

Structurally, cans are sufficiently strong to withstand static, compressive and dynamic stresses. Thus, corrugated board might be considered overpackaging if adequate alternatives could be found. Shrink film makers are hoping that their products might be satisfactory. Not only must technical feasibility be proven, but also the various regulatory agencies must be satisfied. Perhaps the most significant obstacle to be surmounted is economics. Overwrapping equipment and shrink tunnels may be more expensive than semi-automatic casing and case sealing systems. The question of packaging material costs has not yet been adequately answered.

Shrink casing is mentioned only as another

food use of flexible packaging materials. It represents a potentially large use for this family of packaging materials.

## Flexpack

Among the names applied to heat sterilized flexible packages have been flexpack and flexican. These are intended to be used in place of rigid metal cans by military personnel. A commercial objective would be the ability to heat-sterilize foods with reduced thermal damage to the food itself by economically altering the surface to volume ratio of the package.

Canning is a process of preserving foods, normally subject to microbiological deterioration and oxidation, rendering them free of microorganisms until ready for use. The metal can offers a rigid, impervious and commercially inert package which can be sealed to exclude contamination and atmospheric air which could oxidize desirable components. Destruction of microorganisms is effected by heating, with low acid foods by temperatures above that of boiling water. With high acid foods (i.e., less than about pH 4.5), 212°F is an adequate sterilization temperature.

The metal can well withstand both temperature and internal pressure. Many low acid foods contain particulate solid mass which must be thoroughly heated to the required temperature. To assure that sterilization temperatures can be achieved, there must be a thermal gradient from surface to the interior. The temperature differential leads to product cooking and in practice, some overcooking. Alteration of the can shape from cylindrical to flat would reduce the amount of heat required to effect sterilization and would thus reduce thermal damage to the food.

A flat shape could, of course, be attained with a metal can, but the cost of the can, now in the range of $0.04-0.07, would be significantly increased for the volume of contents contained. A flat package of similar dimensions, fabricated from flexible materials, would be considerably less expensive. A flat package would also occupy less space in shipment and on retail shelves because the voids between circular surfaces would be eliminated. Flat packages would also offer the opportunity to increase the graphic display area.

For military individual ration packages, the metal can is awkward to carry. It is bulky in a pocket or sack, and, in the event of a fall by the soldier, both can and man are subject to damage. A durable flexible package would not injure the person carrying it, and could be less susceptible to damage itself.[39]

One important military objective is to assure that the package remains intact throughout the many and various stresses it undergoes prior to use. Because low acid foods are involved, any package failure could lead to toxin formation within the unit. Thus both body walls and seals must be void-free. Minute leaks could allow ingress of microorganisms which could cause spoilage. Assurance of package intactness has been the major obstacle limiting the adoption of the system which has been under development.

The government's goal for an acceptable level of defects is 0.01%, or one defect in ten thousand. At present the best that is claimed to have been achieved is 0.5% in pilot production, but tests indicate more leakers in the field.

A logical methodology that could be applied to the objective would be aseptic processing and packaging. U. S. Army Natick Laboratories spokesmen, however, have equated aseptic processing to high-temperature, short-time heating for particulate materials. "In addition, aseptic processing has not been accepted across the board in the food industry. We would like to see greater acceptance before we look at it. Still another consideration is that there is still the seal contamination problem. We will still have to fill a food material into a package. We have considered aseptic; we have not completely ruled it out."[25]

Particulate foods are sterilized outside of packages for some soups and in the Flash-18 process for meat canning. Many boil-in-bag frozen foods are being packed and sealed under commercial conditions with vacuumization of contents; most frankfurters are packaged under vacuum in flexible materials.

The flexpack system involves use of preformed bags which are filled with food, evacuated and heat sealed. The sealed packages are then racked so that ballooning from internal pressure buildup is minimized. Packages are heated to temperatures above 212°F with external pressure being applied but before the temperature of the contents has been reduced to limit internal swelling on cool-down.

The processing sequence indicates several stresses on the package not ordinarily experienced by flexible packages. Filling, of course, could contaminate the seal area and reduce its effectiveness. During heating, the temperature of the contents approaches that of the softening point of the sealant plastic, generally polyethylene. Internal pressures build up and stress both the body wall and seal areas. The package must be evacuated to eliminate the possibility of oxidation of contents. Even infinitesimal leaks could allow access of oxygen which would deteriorate the contents. Degradation of plastic laminants could increase permeability.

Subsequent field handling can abrade and stress the packaging material and create the unwanted leaks.

A large number of different flexible materials have been evaluated by the U. S. Army Natick Laboratories. Almost all of the laminations have included aluminum foil which is the only totally impermeable commercial flexible packaging material. Gages of aluminum foil are heavy, i.e., over 0.0005", to assure against pinholing which had been common to thinner gages. The foil must be protected against abrasion, puncture and tearing by a tough external barrier. Although paper is used in commercial products because of its low cost, polyester is a far stronger (and more expensive) material and has been the external barrier in many of the test packages.

The most commonly used sealant is a retortable polyethylene. Attempts have been made to use Nylon-11 which is claimed to seal through contaminants better than polyolefins. There have apparently been some experiments with the use of ionomer and rubber modified high density polyethylene as a sealant.

Published reports indicate that all testing has been done using preformed bags of the three side face-to-face seal type. Flat bag formation, of course, allows for two sides to be well sealed. According to U. S. Army Natick Laboratories, " . . . we steer shy of four-edge seal, the reason being that if you make a machine-made bag in the flat stock and then fill it, the two end seals are perfectly aligned and flat. But since it is filled with something, the pouch is distorted to receive the product or fill. When vacuum is applied, the pouch collapses around

the product or fill. This distorts the side seals and they are no longer flat or parallel one to another.

"Actually, the side seals are under stress and will have wrinkles. These serve as stress concentration where flexing and abrasion will be concentrated . . ."[25]

Evidently, there is no satisfaction that the desired defect level can be achieved using flat bags. Yet, much of the present effort is focusing upon the achievement of effective seals in continuous form, fill and seal equipment.

A question was posed as to the reasons for the holdup on flexpack. The answer from U. S. Army Natick Laboratories: "First, economic; and second, the caution we have come to expect all American food packers to exercise in introducing a new product.

"At the risk of oversimplification, I think that both problems can be answered in one word. And that is automation. We can talk about microbial penetration. We can talk about the lack of a perfect oxygen barrier. We can talk about the stress of seals. We can talk about the importance of the overwrap. All of these things are true. But yet the statistical data that exist indicate that the vast majority of all imperfections had to do with the top seal closure or the final closure. We feel that this is the key to the problem."[25]

The country's food industries, with few exceptions, have not led the way in packaging innovation. They have depended upon their suppliers. To expect a food canner to develop a total flexpack system would be counter to experience. Can companies have heavy investments in rigid can making equipment, yet they have been in the forefront of flexpack development, both as a defensive gesture and as a potential opportunity for the future. Most flexible packaging material suppliers do not have the technological resources to develop this system.

Some have suggested that government requirements are too stringent for commercial needs. Others have suggested that striving for perfection in these packages has, in fact, created a moving target rather than a feasible technology.

There could be considerable subjective speculation as to why this program has not become commercial after so many years and so much manpower has been expended. Flexible packaging material suppliers have implied that their tests have proven satisfactory. But vendors have been known to be over-enthusiastic in their claims. As expressed above, even when there is a substantial user demand, suppliers are somewhat hesitant about making developmental investments. When a market need is not openly expressed, courageous management is required to make the investment.

There is no question that much information has been developed on the topic on behalf of the government. The published material, however, does not supply sufficient data to form a definitive critical review of the present or potential status of this undertaking.

Meanwhile, there have been two publicized commercial ventures; one in the United States, one in Europe. The former involved what would be considered a relatively easy product to handle in a flexpack system: sauerkraut. One of the major aluminum foil manufacturers, Reynolds Metals Co., was apparently partly responsible for technological development of a flexible aluminum foil lamination package which was filled with sauerkraut and heat sterilized at boiling water temperature. Canned sauerkraut suffers from relatively poor heat transfer because of the solid pack, and, as a result, tends to overcook. By placement in a flat pillow pouch, it was felt that heat transfer would be less of a problem and a more superior product would result. Unfortunately, the published material was in the form of publicity releases rather than hard data subject to technical interpretation. This product has been withdrawn from the market.[11]

The European entry was also in the form of publicity releases rather than technical publications. The products involved are processed meat, i.e., frankfurters and sausages.[54]

Produced in Denmark for marketing in England, the processed meats are vacuum packaged in a polyester/aluminum foil/polyethylene lamination which is claimed to be heat sterilized after sealing. The claim is that the product is stable at room temperature for several months, perhaps in much the same manner as canned Vienna sausages would retain their processed quality. According to the available information, the flexible packages are suggested for boil-in-bag preparation by the consumer.

## Multipacking of Cans

Shrink packing of cans, in lieu of packaging in corrugated shipping cases, is too far removed from flexible packaging to be included here. Shrink wrapping of beer cans in consumer multipacks, on the other hand, is an integral part of beer can packaging.

There have been several commercial efforts to introduce shrink wrapping to beer can six-packaging. One, by Reynolds Metals Co., has had some limited success. The other, by Du-Pont, has been under very careful test for about five years and, in 1969, was announced to be commercially available.

A brief review of beer can six-packing might serve to highlight some of the problems which have hampered the system. Beer canning is high speed packaging, commonly ranging from 900-1200 per minute. Well over three-fourths of all beer cans are packaged for marketing in six-packs. There is no space in brewery bottle shops for beer cans awaiting six-packing. Thus, six (and eight) packaging must be coupled directly to the end of the canning line.

To meet this need, paperboard converters, in conjunction with equipment manufacturers, developed continuous motion in-line wraparound banders which use die-cut paperboard blanks. The resulting paperboard created an opaque package which hid the cans brewers have so meticulously designed.

Conex Division of Illinois Tool Works developed a die-cut extruded rigid polyethylene structure which fits around the upper can chimes to hold the cans in a fixed, six-pack configuration. This package provided total visibility for all the cans and, more important, was far less expensive than paperboard. Further, the supplier developed a continuous motion in-line machine which utilized roll-stock die-cut polyethylene and which was apparently more efficient than paperboard equipment.

Brewers found that the Conex Hi-Cone® system did not afford them the opportunity of adding a sales message. Further, the system was so sound that it was difficult for consumers to identify the fact that the cans were in six-packs. The ends of the cans, now almost all of which have convenience tops, are exposed to dust, dirt and debris. The advantages far outweigh the disadvantages, and the Hi-Cone® carrier has captured a major share of the business.

Brewers who were seeking a merchandising advantage examined bundling in transparent film. Cans could not hold together neatly and without jostling in a loose overwrap such as paper. Internal movement within a wrap was found to abrade at chime against chime contact points. If cans slip vertically so that chime to body contact is made, potential for damage is increased because only one thickness of metal is present, and tin-plated steel (or tin-free steel) can be broken. An overwrapping system thus had to hold the cans tightly together to prevent movement against each other.

Shrink films, as materials, had the desired properties. They could be bundled around the cans, sealed and shrunk to hold the cans tightly against each other. The per cent shrinkage could be such that, even with some film relaxation, the cans would still be tightly held. Shrink films also are true thermoplastics not subject to significant changes from environmental moisture and generally fairly durable with little tendency to propagate tears. Most shrinkable plastics such as PVC and oriented polypropylenes, however, are soft films and not nearly as machinable as stiff films.

Shrink film bundling provides a transparent wrap although with some materials, such as polyethylene, there is cloudiness. All show sufficient wrinkling at stress junctures to clearly demonstrate their presence. Once the wrap is torn open by the consumer, there is a tendency for all of the cans to tumble from the remainder of the bundle. Shrink films are not readily decorated because printing would be distorted by the shrinking process. This is overcome by adding a strip of printed material beneath the shrink film to carry the message. Carrying holes are burned into the film after shrinking to strengthen the edges of the holes and minimize the possibility of film tearing.

All of the above problems are minor; two serious problems exist. Can chimes abrade through shrink films in six-pack to six-pack contact in shipping trays. Secondly, film characteristics have made high-speed machining (needed on beer canning lines) difficult.

The first problem is not easy to overcome with three-piece cans: double seams protrude away from the can body and create lines of contact. This has almost precluded shrink film from conventional three-piece can bundling. The two-piece aluminum can, now being used for about 10% of all beer cans, has a necked-in top. The double seam is thus recessed and flush with the can body. Chime to chime contact is minimized if not eliminated. Further, can bodies are tight against each other with relatively little possibility of slippage and consequent loosening. Much shrink bundling, thus, has been with two-piece aluminum cans. Development of the top and bottom neck-in, three-piece, tin-free steel can has not proceeded far enough for this package to be a factor.

The problem of machinability has not been totally resolved. It is inherently more difficult to handle a large sheet of soft film, wrap it totally around six cans with overlap at each end, and seal it on the bottom than it is to snap a paperboard band around or snap plastic rings around chimes. Further, a printed film or paperboard piece is placed under the shrink film, finger holes must be made, and, finally, at a subsequent station, heat must be applied to shrink the film around the cans—all of this at 200 six-packs per minute.

Reynolds Metals Co., in conjunction with George Meyer, a machine company with considerable beer packaging experience, developed equipment to perform these tasks from roll stock PVC. The system works satisfactorily for several brewers. The equipment size alone is considerably above that used for the other two basic six-packing systems.

From the problems, it is easy to understand DuPont's hesitancy to move its Clean Pak® system into full-scale production operations. The machine developer with DuPont is R. A. Jones Co., a highly respected custom packaging equipment designer and builder with experience in tray systems for beer packaging. DuPont's film is its Clysar® heat shrinkable polyethylene.

# CEREALS

## Summary

The cereal business may be segmented by product: hot cereal, instantized hot cereals, ready-to-eat cereals, enriched and sweetened ready-to-eat cereals. The original shelf stability of hot and ready-to-eat cereals led to concentration of manufacture among a handful of large companies. Distribution is from warehouses to retail outlets and so is out of direct control of manufacturers. Marketing of a substantial share of product is child-oriented and so undergoes frequent product and graphics changes.

Hot cereals today are in basic package constructions created over 60 years ago: simple lithographed cartons. Instantized cereals do not require a high degree of protection against moisture or fat seepage and thus are sufficiently packaged in unlined chipboard.

Precooked and flavored precooked hot cereals are composed of hydrated starch, and thus require moisture and odor protection. By portion packaging this product, protection is afforded through envelope pouching using paper/polyethylene fabricated on horizontal form, fill and seal equipment. Multiple filling of pouches into retail cartons is done on equipment which is an integral part of high speed pouching machines.

Almost all ready-to-eat cereals are expanded and toasted grains susceptible to moisture damage of texture and to fat exudation. Most packaging is in a larger multiple portion bulk package requiring reclosure in the home. A large number of individual portion packages are marketed to HRI outlets in case lots and in multiples for the retail trade. Virtually all ready-to-eat cereal packaging uses lined paperboard cartons with the carton acting as the structure and straight-side form for stacking and providing a large display face.

The inner liner imparts moisture and fat protection and has the ability to reclose. Unsweetened cereals generally use waxed, single-ply, cereal grade, glassine liners. Sweetened, flavored cereals employ a variety of materials ranging from overwaxed glassine laminates to foil/wax/tissue to triplex (glassine/wax/aluminum foil/wax/glassine) to overwaxed, one-side saran emulsion coated glassine. The sugar on the product surface is hygroscopic and demands moisture protection.

All lined paperboard packaging is preformed on double package maker equipment. Unsweetened cereals use foldover closures; sweetened cereals requiring added moisture protection are often heat sealed for closure.

A small quantity of dry cereal is portion packaged in thermoformed tubs which are closed with a paper/foil lamination, heat sealed to the lip.

## Introduction

The breakfast cereal industry has traditionally been divided by product category; hot and ready-to-eat, sweetened, flavored, enriched with vitamins and minerals, and synthetically sweetened. Individual portion packaging has become a small but significant part of the market.

Products for hot preparation originally were basically refined and reduced particle size grains such as oats, wheat, or corn. Most products have been improved by processing or by the addition of chemicals to alter their cooking characteristics. These improvements have been aimed at reducing cooking time. They remain products to which liquid is added to hydrate and swell the starch components and create a hot food, usually consumed at breakfast.

More sophisticated techniques have been employed to reduce the cooking time to the "instant," i.e., one-minute cook or mere addition of hot liquid. These have included, in some instances, precooking to hydrate the starch or pretreatment with enzymes. Unaltered grain cereals are fairly stable to moisture; those which have been slightly processed are also quite resistant to moisture although not completely stable. Prehydrated cereals, on the other hand, are relatively susceptible to damage as a result of water pick-up.

A number of hot cereal producers have expanded their markets (or retained their market share in a declining market) by adding flavors to the basic product.

Ready-to-eat cereals are generally processed by addition of flavoring agents, precooking and subsequent aeration and drying to create puffed, crisp products. Because the grain is cooked prior to expansion, several grains and

33

grain derivatives, such as flour, can be mixed to obtain various flavor and texture effects. By extruding and expanding through different dies and with varying toasting and heating temperatures, a variety of different products may be obtained. Thus, whole grain rice may become puffed or crisped; puffed, shredded and wheat flakes may result; puffed and toasted oat and wheat or oat and corn may be obtained. The products may be shaped into animals or letters of the alphabet. They may be made large or small depending on the desired end result. Ready-to-eat cereals are generally designed to remain crisp.

The technology of cold cereal manufacture has been expanded by the incorporation of flavors into the formulation, and by the addition of sweeteners, colors and flavors on the surface. The basic sweetener is sugar syrup. The added flavorings are cocoa in syrup form to impart chocolate flavor, fruit-flavored syrups and other complementary components.

Some cereals are sweetened using synthetic sweeteners in natural gum bases. These may also have flavor additives with the sweeteners. Ready-to-eat cereals are also altered by the inclusion of other products such as textural and flavor additives. Examples include candy, such as marshmallow, and dried fruit, such as raisins, or freeze dried bananas, strawberries and blueberries.

The basic shelf stability of the resulting products means that they may be manufactured at central locations and distributed throughout the country. The cereal business is almost exclusively in the hands of a small number of large companies, all of whom distribute nationally. None is exclusively in the breakfast cereal business, but each has a number of products which constitute a significant part of the line. Most have a broad line of other breakfast cereals competing against each other.

Although the leading products in each product category are oriented toward no special demographic group, newer products are believed to be marketed toward children. Both product and packaging provide evidence to substantiate this belief. This type of market orientation leads a constantly changing group of consumers who enter the market as youngsters and leave as teen-agers or as adults. It is also a market which can be swayed by external influences such as advertising or astronautics. It is thus a market which demands new products and new packaging (at least graphics) at a prodigious rate.

The market is not easy to penetrate. Cereal manufacturers do not have their own driver-salesmen. This, of course, means less control over product and its merchandising, and results in a need for higher marketing effort.

The business has not been penetrated by new companies in several decades. Neither is there more than a handful of small companies to manufacture on a private label or regional basis. This means that the giants compete against each other for a market which has been assaulted on the one hand by a trend to skip breakfast and on the other hand by instant liquid breakfast.

Although the relative positions of the leading products in each category have remained about the same for many years, their shares have been eroded by the proliferation of new products.

**Packaging Requirements**

Except for the bowl-ready instant products, cereals do not have exacting packaging requirements. They are dry and thus are not subject to microbiological deterioration. Hot cereals absorb moisture only with difficulty and so require little moisture protection. Fat contents of other than the whole grain cereals have been reduced considerably, and so rancidity from exposure to air is not a major problem (although it is a factor of which to remain cognizant). Many flavoring agents are stable, and little protection is required to retain them. Vitamins used for enrichment do not present any problems if the cereals are kept reasonably dry.

Many cereals have been and continue to be packaged satisfactorily in containers designed before the turn of the century.

The basic hot cereals require little moisture protection. Since they are made from starchy fractions of grain, the germ or oil-containing fraction has been removed, and they are not subject to fat alterations from exposure to air. There is no particular problem of fat exudation through the packaging. Some cereals are subject to insect infestation and extra protection is required to keep moths and beetles from penetrating the packages at a weak spot, such as a corner.

Bowl-ready precooked cereals contain hydrated starch which is designed to accept moisture readily and which, therefore, requires water vapor protection. The incorporation of flavors, such as cocoa or cinnamon, leads to the need for an odor barrier supplementing the moisture barrier.

Ready-to-eat cereals in contrast are hard and often fragile. They can scratch and be abraded and crushed if seriously abused. The light weight and density allows them to flow rather easily and move inside containers with only minimum surface powdering and breaking. Even external forces up to a point only succeed in moving aside the cereal rather than physically damaging it. There must, however, be some structural protection.

Unsweetened ready-to-eat cereals require some moisture protection because slow absorption of water can lead to loss of crispness and a toughening of texture. This reaction imparts unwanted chewiness. Most moisture transfer reactions are reversible so that after absorption, the product can be crisped by placement in a low water vapor environment—all relatively slowly.

Ready-to-eat cereals may be produced from a variety of degenerated grains and grain flours with varying quantities of fat remaining. While the bulk of the fat subject to rancidity has been removed prior to processing, there is a long-term possibility of rancidity and short-term potential from fat flow into packaging as a result of temperature fluctuations. The need to protect these cereals from odor contamination of cylinder board cartons in which they are contained also exists. As with crackers and cookies, ready-to-eat cereal products often need to breathe through the package.

Sweetened cereals present a more difficult moisture problem because dried sugar syrup is an active moisture absorber when the vapor pressure exceeds the equilibrium. Moisture absorption leads to partial liquefaction and surface stickiness on one part of the spectrum, and binding of the pieces together at the other end of the problem range.

All of this points to a requirement for moisture protection. Some advocate total sealing and others a good barrier without the need for hermetic sealing. Regardless, there is the need for reclosure in the larger sizes. Once the package is opened and only part of the cereal is consumed, the remainder must be protected against the hostile and moist environments of breakfast table, kitchen and pantry.

Synthetically sweetened dry cereals do not contain additives which readily absorb moisture, and so water vapor protection requirements for these would be about the same as for unsweetened cereals. Those with added nutrients, however, (and most cereals are fortified with certain vitamins and minerals) may contaminate other products with the flavor of the vitamin.

Cereals with inclusions such as raisins, nuts and candies do not have major problems of moisture uptake. The cereals themselves absorb water at a more rapid rate than the inclusions.

Cereals with freeze-dried fruit (all have been withdrawn from the market) experienced just the reverse. The porosity and low moisture content of the dried fruit structure led to extremely rapid rates of water absorption by the fruit. If the moisture available was insufficient for total reconstitution the moisture distributed itself throughout the fruit and led to toughening. Moisture absorption in freeze-dried fruit was not reversible. Once absorbed, the moisture was removable but through conventional evaporation mechanisms rather than by sublimination, which changed the texture for the worse.

Freeze-dried fruit is very fragile material subject to breakage and powdering. The powder increased the surface area of dried product which could also absorb water.

The ready-to-eat cereal marketing situation referred to above indicates a need for an ability of the packager to change graphics rapidly and smoothly.

With a strong trend toward individuality in products, there is an increasing movement towards portion packaging of both hot and ready-to-eat cereals. Some of these packages are intended to act as the consuming container. With this trend has come a need for more multipacking of the individual packages.

Cost is a relative matter. Cereal is not a commodity and so is not priced directly against competition. Neither does cereal have a fixed retail price. On the other hand, cereals are not specialty products requiring sophisticated packaging, and so packaging costs cannot be excessive.

There is a distinct dichotomy of packaging in cereals with a never-ending investment in graphic changes and a pressure for reduced cost of materials. Fortunately, there are few pressures for proliferation of sizes, and, probably due to legislation, this situation will remain.

Little has been stated or implied concerning the application of flexible packaging, per se, to cereal packaging. The requirements might be met by a number of packaging systems; in the case of cereals, the requirements do not specifically define flexible packaging.

### Packaging

Old-line cereals not requiring much protection are packaged in linerless chipboard cartons. The traditional paperboard cylinder with a lithographed paper label is a part of America which has been often copied by competition and accepted as synonymous with the oatmeal company which developed it. It continues to be used successfully. Farina and whole wheat hot cereals in rectangular paperboard cartons with printed double wound overlabels (to protect against insect infestation) are close packaging relatives of the cylinder. These packages are inexpensive, readily decorated, stack well on top of each other (and in the case of the rectangular containers, in blocks) and might even be considered overpackaging in the harsh light of today.

Instant, bowl-ready cereals which require moisture protection are in one of the least expensive material constructions which can effect that protection: a paper/polyethylene pouch. The pouch contains an individual portion and is easy to open and dispense. Polyethylene extrusion on the interior provides an inexpensive moisture barrier which also acts as a heat sealant. MG paper could be used if only moisture protection were required. But since some aroma barrier is needed, pouch paper or glassine is used as the substrate. The pouch form (three- or four-side, face-to-face seal) is one usually needed for moisture protection. Pouches are multipacked in paperboard cartons on edge, the configuration which potentially imparts the greatest damage to seals and areas adjacent to seals.[37]

Ready-to-eat cereals in family-size packages are almost all packaged in DPM, lined, rectangular paperboard cartons. The liners provide the barrier protection and the paperboard provides structure, form, stackability and good display facing.

Liners for unsweetened cereals belong to the same family of waxed and laminated glassines as the cookie and cracker package liners. For cereals, however, greaseproofness is a lesser problem and a more open sheet may be used. Furthermore, moisture protection may be considered a lesser problem, and so the total amount incorporated in the sheet may be less. Wax, of course, acts to retard moisture passage, with impermeability increasing as a function of weight per ream. The glassine acts as the fat and odor barrier. Generally, two-side bleached or amber waxed glassine is used, amber being very slightly less expensive.

Waxed glassine liners are heat-sealed on the long and bottom seams and foldover closed on the top after filling. Waxed glassine has sufficient stiffness for foldover closure.

Shredded wheat is now individually packaged (a reuse convenience as well as moisture protection) in paper/plastic heat sealed face-to-face.

There are some differing philosophies on the means for fulfilling packaging requirements for sweetened cereals. Some employ wax laminated, overwaxed glassine with double foldover closure, with their tests indicating that this is satisfactory for both large and small packages (with their larger surface-to-volume ratios). Wax laminated, overwaxed, glassine liners may also be heat sealed and folded over for added protection.

Other barriers are employed for dry, ready-to-eat cereal protection. Aluminum foil wax laminated to tissue is used on some individual portion packages. Since the package may be used as an eating container to hold liquid milk, the foil on the outside serves to retain liquid during use. Wax on the interior acts as a sealant because a total seal is required to assure against leakage. The wax laminant is a moisture barrier.

Although a similar material could be employed for larger packages, one construction which is employed is a triplex, glassine/wax/aluminum foil/wax/glassine. The glassine can be inexpensive cereal grade because its principal function is to protect the foil which provides the major barrier. Aluminum foil is now used in 0.003" and even 0.0025" gages instead of the former 0.00035". The small pinhole level in these thicknesses is readily overcome by the microcrystalline laminating wax and provides an effective moisture barrier. Use of two wax layers leads to 8-10 lb of wax per ream, sufficient to provide the required moisture protection when combined with aluminum foil.

Glassine on the inside protects the foil against abrasion from cereal, is a fat and odor barrier, and also separates the product from the wax, reducing the possibility of waxy flavors being imparted to the cereal. Foil's dead fold characteristics allow for foldover reclosure.

Another construction is aluminum foil/wax/ strike-through tissue.

The fact that some cereal manufacturers use no foil laminations for almost identical products is indicative of the questions that could be raised concerning the need for these constructions (which are, incidentally, among the least expensive aluminum foil laminations). Aluminum foil construction prices have been declining in the last five years until they are only slightly above those of other laminations not incorporating foil.

Into the area of sugar sweetened cereal liners has come what is sometimes referred to as an intermediate barrier: overwaxed one-side saran coated glassine or paper. Saran is not as impermeable as aluminum foil but is the best of the commercially available barrier coatings.

The data which became the input to the decision on employment of one material or another are not proprietary, but neither are they published. Information on permeability in the flat is available for some typical structures, but typical does not necessarily mean that any specific sheet is being used. Further, and most significant, permeability in the flat may not in any way reflect the actual barrier properties in package form. Flat WVTR is but a relative indicator; creasing causes alterations and package formation brings in the highly variable parameters. Although almost all dry cereal packages are formed on double package makers, there are variables, such as how the seals are formed. Unless completely comparable tests are performed, reports on the relative capabilities of flexible structures cannot be evaluated.

Such information is really not of broad enough consequence to warrant publication. There is little of potential interest in a recitation of moisture pickup by various ready-to-eat cereals in different structures. It would, however, appear from observation of the several different constructions used for the same cereals, that there are at the very least some differences in interpretation of data. Perhaps there are differences in experimental procedure which lead to this difference. Since it would appear that one or more packagers are overpaying or underpackaging, a definite comparative study by a supplier or an outside agency might be profitable to cereal manufacturers.

All of the above commentary is based on continuance of the employment of the double package maker as a cereal package for both family and portion packages.

Portion packaging of ready-to-eat cereals has been in both lined paperboard cartons and in sealed thermoformed tubs. One of the major cereal packagers packs a limited amount of ready-to-eat cereal in vacuum formed vinyl tubs which are sealed around the flat lip with peelable, printed, paper/aluminum foil lamination. The opaque white vinyl tub is a reasonably good moisture barrier, and also holds milk added by consumers.

DPM portion pack cartons are multipacked in tens, and other multiples, for retail sale. For HRI and some retail use, the cartons are bulk cased. Multipacking is by overwrapping in printed cellophane or polyethylene, with or without a printed paperboard tray. The package is conventional overwrap bundled, using transparent film and printing to show the contents or a paperboard tray to express the message of many packages in a single unit.

**Equipment**

As indicated, the major equipment used in breakfast cereal packaging is the double package maker by Pneumatic Scale Corp. Those comments made for the use of the DPM for cookies are equally applicable to the employment of this equipment for ready-to-eat cereals.

It would be most difficult for DPM machinery to use soft plastic films as carton liners which might afford a cost savings. DPM's depend on sheet stiffness for formation. Bag-in-box systems described above are, of course, more expensive than DPM from both equipment and material standpoints.

There has been considerable thought and testing given to the use of linerless paperboard cartons. Some have been placed on the market, having been proven technically feasible up to a point. Linerless cartons can be made moisture-proof (at least to meet the requirements of dry cereals) until opening. Reclosure is a problem with linerless paperboard cartons, although the packages in the marketplace may be acceptable.

It is difficult to demonstrate in experimental and field tests the equivalence or superiority of linerless cartons over lined packages. It may be even more difficult to incorporate protection into a paperboard carton at a lower cost than that of a lined carton. Regardless of the presence or absence of a liner, the DPM remains the equipment best suited to the package.

From the tremendous investment by cereal makers in DPM's, it appears highly unlikely that these machines will be replaced for existing products, especially in view of the fact that no apparently suitable package concept is being developed. This does not imply that there will be any substantial increase in use of this concept for new cereal products which would expand the market. Ready-to-eat cereals, which are extensions of current lines to replace those which will fade from the market, will doubtless be in lined paperboard cartons. Wholly new breakfast cereal products will be needed for innovation into new packaging concepts. The use of pouch packaging for flavored instant hot cereals is an example of new packaging for new products.

It might be expected that packages serving as eating containers will become more a part of the ready-to-eat cereal packaging scene in the future. Changes in flexible packaging, as a result, would be difficult to forecast.

The possibility exists for stand-up bags made from flexible materials. Fully automatic equipment to form, fill and seal flat bottom square sided bags from roll stock is on the market. Whether such equipment will be applied to cereal packaging is speculation at this time.

The latest major commercial cereal packaging method is pouch packing of instant hot cereals. Because several contract packagers have automatic equipment for manufacturing and filling this package type from roll stock, the initial pilot work must have by-passed pre-formed bags. Three- and four-side envelope pouches may be made on several different makes of equipment, but cereal packaging is most often performed on horizontal form, fill and seal units. Bartelt continuous or intermittent motion equipment is probably the most widely used in the United States. Bartelt equipment operates at speeds above 60 pm for intermittent motion and up to 400 pm for continuous motion. Bartelt equipment can continuously couple pouch making with single or multiple cartoning, using sleeve-type paperboard cartons. The Canadian Delamere and Williams equipment is probably second most popular.

Cereal pouch packaging represents a logical transfer of a packaging concept from another successful area, dry soup mixes. The total packaging system was available and thus did not require development of material, primary package or secondary packaging equipment.

---

# TABLE 1

# POSSIBLE COMPONENT LAYERS

# IN FLEXIBLE LAMINATES IN

# PACKAGING

Polyethylenes
Polypropylenes
Polyvinylidenechlorides
Polyvinyl acetates
Polyvinyl alcohols
Polyesters
Polycarbonates
Polyurethanes
Polystyrenes
Polyallomers
Phenoxies
Ethylene-vinyl acetate copolymers
Ethylene-ethyl acrylate copolymers
Fluoro and chloro-fluoro hydrocarbon polymers
Ionomeric copolymers
Vinyl copolymers
Block and graft copolymers
Papers
Paper-like webs of mixed cellulose and plastics

Papers made of plastics
Bonded fibre fabrics
Cloths and scrims
Spun bonded fabrics
Regenerated cellulose films
Cellulose esters
Cellulose ethers
Rubber hydrochloride
Chlorinated rubbers
Chlorinated polyolefins
Natural and synthetic rubbers
Natural and synthetic waxes
Natural and synthetic bitumens and asphalts
Natural and synthetic resins
Adhesives of all types
Prime, key, bond or sub coats
Latex bound mineral coatings
Aluminum and steel foils
Deposited metal layers.

Source: Mr. E. V. Southam, Director, Dickinson Robinson Group Ltd., Bristol, England, *Packaging Technology*.

# COFFEE

## Summary

Coffee is generally marketed in the form of roasted and ground coffee for brewing, spray dried instant, and freeze dried instant. Roasted and ground is sold in two markets, retail, and hotel, restaurant and institutional. With the exception of portion packaging of decaffeinated soluble coffee, instant coffee is absent from the HRI market.

Roasted and ground coffee is very susceptible to loss of aroma by volatilization, oxidation of flavors and oils, and acceleration of deteriorative reactions by increasing moisture. Roasted and ground coffee must, therefore, be packaged in the absence of oxygen and with an effective barrier against gain or loss of gaseous materials. At the retail level cans with polyethylene reclosure lids are almost universally used. Because turnover is so rapid in the HRI market, lined paper bags have been used. To afford greater operator flexibility, however, vacuum packaging of roasted and ground coffee in flexible packages is under intensive development. The concept of bag-in-box has been tried for retail packaging, using polyester/aluminum foil/polyethylene pouches. Other combinations incorporating polyethylene as a sealant, saran as the gas barrier, and polyester for durability have been employed for pouches not in paperboard cartons.

Portion packages of decaffeinated instant coffee are packaged in four-side sealed paper/aluminum foil/polyethylene pouches.

## Introduction

Two markets exist for coffee's three products: retail and HRI (hotel, restaurant and institutional).

The three products are roasted and ground beans, the familiar product packaged in vacuum cans; spray dried soluble, powder-like-instant coffee sold in glass jars; and the well-advertised freeze-dried coffee which might be classified as a sub-category of soluble. A small amount of unground coffee beans is sold on the retail level, and a small amount is roasted for grinding and packaging in retail stores. Many of the spray dried solubles are being agglomerated, a process which makes them more soluble, and which gives the appearance of the freeze dried product (although the latter could be reduced in particle size to a powder). Further subdivision could be made—roasted and ground coffee (R & G), compressed rings, decaffeinated, and even flavored.

The home market uses over 35% soluble coffee with the balance brewed from roasted and ground. Because the typical coffee-drinking family contains only a few persons per consuming period (e.g., breakfast), only a few ounces of roasted and ground coffee are added to the coffee brewing device at any time. R & G coffee must, therefore, be stored for a day, several days, a week, or perhaps even longer. Once the package is opened, the ground coffee loses flavor rapidly, a deterioration which may be slowed somewhat by refrigerated storage.

The situation with soluble coffee, regardless of the drying process, is quite similar: the desired portion is removed from the container and consumed. The remainder is stored in its original jar until needed, the duration being days or weeks.

The converse is true in the HRI trade. Coffee is a popular menu item, with some establishments selling the beverage as their main product. HRI operations purchase R & G coffee in large multiples of sealed units. Each unit (or multiple) contains enough roasted and ground coffee to make one pot, or a larger quantity of hot beverage. The brew is quickly sold, within minutes or hours, and the coffee maker is recharged with R & G coffee to make more brew. Because coffee is such a steady item, the amount which will be consumed is predictable. Only the amount of R & G coffee needed to supply the expected demand is purchased. Each unit remains sealed until used. Turnover is fast, so that there is little chance for even the packaged coffee to undergo changes.

Instant coffee in bulk is not used to any extent by the HRI trade. Some individual portion packages of special instant coffee, such as decaffeinated to be brewed at the table or by the consumer, are sold.

Coffee consumption, despite appearances, has not increased on a per capita basis for many years. Instant coffee, first really promoted 20 years ago, (although it had been available for many years earlier) represented a marketing innovation. Freeze-dried coffee is the result of a major investment to produce incremental product improvement. It must be considered a marketing innovation which has not significantly increased the per capita consumption of coffee beverage.

Roasted and ground coffee underwent no product changes for many years, with much of its marketing innovation in evolutionary packaging changes.

With the legislatively imposed demise of several of the coffee industry's favorite point-of-purchase promotional activities, and with advertising channels saturated, the most obvious untapped area for moving R & G coffee may be in packaging. Some attempts to introduce R & G coffee into the retail market in bag-in-box combinations have been made.

There have also been studies to determine the feasibility of soluble coffee for the HRI market.

The coffee industry is dominated by a limited number of large companies, most of whom have interests in other areas of the food business. The largest company in the coffee business holds a little less than half of the retail and HRI R & G trade, and the retail spray dried and freeze dried instant coffee markets. The secondary companies in each of the product classes are large both in total size and in the coffee business. Those in the R & G industry also have spray dried instants, and sell to the HRI trades. This is not true of some of the larger factors in the soluble coffee business, several of whom do not operate in the roasted and ground field. They are extractors, packagers, and marketers in the dry grocery trade.

Vigorous competition exists among the larger companies. A few of the very large firms are regional, although some of their regions are very large. There are also some strong local roasters and grinders serving large metropolitan marketing areas. Private label coffee is common at the retail level. A number of contract soluble coffee houses provide private label instants, now in both spray and freeze dried form.

**Packaging Requirements**

Coffee is consumed in great measure because of its flavor. This flavor, extracted from ground beans on brewing with hot water, is developing in growing, post-harvest processing and roasting. After roasting, which imparts desired flavor for extraction, degradation occurs. Grinding exposes more surface area for extraction and equally to deterioration. In-store and home grinding of the whole beans are methods used to obtain desired flavor and roasted beans.

The consumer purchases preground coffee which must be protected from the environment. Roasted coffee contains aromatics which can be lost as volatiles. It contains flavor components which are oxidizable into undesirable materials. Coffee oil is flavorful, oxidizable and oily. Roasted coffee continues to evolve carbon dioxide gas for some time after roasting. Moisture pickup accelerates biochemical reactions which can cause equally detrimental reactions.

This means that roasted and ground coffee must be totally separated from the environment to retain its initial quality. It should have its own microenvironment to minimize reactions from external influences such as oxygen, light and moisture. Coffee should be within a barrier which minimizes losses of desirable components. Ideally, of course, reduction of temperature would reduce rates of biochemical reaction and volatilization, but this benefit is not sufficiently marked to make the practice commercial except in the home after the package has been opened. Ground coffee is additionally a hard and sharp cornered abrasive material.

Roasted and ground coffee shelf life can be measured in hours, or, at best, days when not completely separated from the ambient atmosphere or stored refrigerated. In a totally sealed package such as a can, initial quality can be retained for several months, but falls off rapidly after opening due to exposure to air.

Instant coffee is the dried extractor brew. The hot water extract is dehydrated by exposure to hot air at atmospheric pressures or in vacuum. The hot air may carry with it some of the volatiles and other desirable constituents.

Spray drying, the most commonly used method, creates a granular powder which contains most of the flavor elements of brewed coffee. Dried coffee is hygroscopic. It also contains volatile aromatic constituents as well as oxidizable oils. Instant coffee is less subject to oxidative processes but more subject to deterioration from moisture absorption than roasted and ground.

In the past few years (perhaps in response to the granular appearance of freeze-dried coffee), spray dried coffee has been agglomerated to give it more rapid rehydration characteristics and the appearance of its more expensive counterpart. The product is manufactured by rewetting and drying the particles which have stuck together into granules with porous and fragile structure. This product has all the requirements of spray dried instant coffee plus the increased tendency to dust and sift.

The effects of drying heat on liquid coffee led to intensive study of the freeze drying process as a method to bring a better quality instant coffee to consumers. Freeze drying is a process in which the extract is frozen and then subjected to very high vacuum so that water vapor is withdrawn by sublimation rather than by evaporation. In this process, very few of the desirable flavor constituents are changed by volatilizing. Freeze drying removes water from a rigid frozen structure and so results in a porous material. Freeze dried coffee is probably more hygroscopic than spray dried instant. Since it contains more desirable flavor components than heat dried plus added-back flavor, there are more constituents subject to storage alterations by volatilization and oxidation.

Instant coffee is a very expensive product. Freeze drying, being a batch process, is expensive, making freeze-dried coffee more costly yet. A product as valuable as this must be protected against pilferage and losses due to package damage in distribution or at the retail level.

These requirements for retention of initial quality do not take into account some of the demands for convenience which have been made. HRI coffee brewing requires methods for quick opening and dispensing. The vending portion of the HRI business has designed a number of different packages for their equipment. Some equipment mixes grounds with hot water; others retain the grounds while hot water passes through; at least one is designed to pour R & G coffee into an extractor from a bulk holder. Several vending machines are designed for soluble coffee from individual packages and others from larger quantity packages.

### Packaging—Roasted and Ground Coffee

Almost all R & G coffee for the retail market is packaged in hermetically sealed and undeniably expensive vacuum cans. Injection molded semi-rigid polyethylene snap caps are on almost all coffee cans now.

One of the more interesting recent test packaging innovations is packing R & G coffee in a porous paper ring-shaped configuration. A ring may be placed over the stem of a basket in a percolator without measuring, and possible spilling. The material is heat sealed from two sheets of flexible material which pass hot water freely while retaining the grounds. A number of rings are packaged in a conventional reclosable vacuum coffee can because the flexible material, of course, offers no protection against environment.

Roasted coffee beans are sold in simple gusseted, flat bottom paper bags. Coffee beans which are ground and packaged at the retail outlet are generally packaged in preformed gusseted flexible bags. These have been of two constructions for many years: paper/glassine and paper/rubber hydrochloride. Neither structure, of course, is an oxygen barrier.

Several attempts to employ European bag-in-box packaging for coffee have been made, and some are on the market. Generally, the inner liner is a preformed pouch made from polyester/aluminum foil/polyethylene. The sealant is medium density polyethylene, because of the potential for interaction of oil with low density polyethylene. The bag is in the paperboard sleeve, and both are opened simultaneously when the sleeve is snapped open. The construction allows drawing of a vacuum after filling, followed by sealing. The paperboard carton acts as a form to shape the bag into a rectangular solid, and helps protect the flexible packaging against physical damage.

Although bag-in-box packages of coffee are cheaper and lighter weight, and stack better than metal cans, they have not enjoyed great success in the American market. There is a large investment in canning equipment which is still in good operating condition. Bag-in-box packaging equipment is more expensive, more complex and more difficult to operate than canning machinery. At about 60 per minute, it is inherently slower than canning machinery which is capable of operation in the several hundred a minute range. Opening and reclosure of flexible packages of coffee is somewhat more difficult than metal cans.

The perpetual labor turnover in HRI operations, however, has presented no such restrictions to this trade. R & G coffee was principally in glassine and rubber hydrochloride lined, paper bags. Paper bags are quick opening and easy to empty into large coffee makers. Lined paper coffee bags, however, have little protective value for the coffee. Shelf life is short.

To increase the shelf life and to provide greater flexibility to both packager and user, greater protection could be built into the package. Oxygen-free packaging in pouches is the major direction being taken in development work with nitrogen-flush packaging appearing to be dominating at this writing. There are some who are adding shelf life by simple sealing under atmospheric conditions in pouch/paper/polyethylene pillow pouches formed on vertical form, fill and seal equipment. Such a system would offer shelf life intermediate between paper bag and vacuum package. At least one system is trying gas flush packaging to eliminate the negative pressure differential of vacuum packaging.

Evacuation of the package during or after filling and sealing requires reliable sealing, demonstrable evidence of package integrity, and

materials which are durable in regular HRI distribution. They need not be so tough as to be difficult to open without utensils. The cost would not be as great as vacuum cans, but would, of course, be greater than lined paper bags. The value of added shelf life would be the return from added packaging cost.

A number of flexible materials have proved satisfactory to hold a controlled atmosphere: the polyester/aluminum foil/polyethylene described above for both coffee and flexican; polyester/saran/polyethylene used for processed meat; polypropylene/saran/polyethylene/and nylon/saran/polyethylene.[49] Some have added an exterior paper for added strength, printability, and stiffness. Within the relatively limited shelf lives demanded by the HRI trade, these laminations in appropriate constructions can provide adequate retention of vacuum. The packages may be envelope type, three- and four-side sealed, pillow-pouch style, or tetrahedral with fin-type long seals. When the vacuum is drawn, the film clings tightly to the product and compacts the coffee. The result is a brick-like block of roasted and ground coffee which can be stacked and handled much like a solid. If the package integrity is violated, it is no longer hard. It may be easily opened by tearing and emptied. More important, shelf life is extended at a cost far less than that of a metal can.[12-14 16 17 21 22]

R & G coffee for vending machines may be in conventional packages which are emptied into the equipment extractors. To illustrate, one machine brews up one at a time by passing hot water through one sealed section of a porous paper magazine. The accordion-pleated magazine is packaged in a protective film in a bulk shipper.

### Packaging—Instant Coffee

Virtually all instant coffee, whether spray or freeze-dried, is packaged in glass jars. One cup portion packages are unusual at the retail level except for promotional samples. The widespread use of flexible packaging for soluble coffee at the HRI level illustrates the technological feasibility of the method.

Little instant coffee is used by the HRI trade, but most eating establishments stock a supply of portion packages of decaffeinated coffee for their customers. Because the demand is limited, relative to that for brewed coffee, turnover is slow. As a result, long shelf life is required.

Instant coffee is generally portion packaged in conventional envelope pouches, constructed from paper/aluminum foil/polyethylene, which provide an adequate moisture barrier and reasonably satisfactory protection against oxygen ingress. Headspace in such a small package is so low that oxidation problems are minimized. Further, instant coffee is not subject to oxidation as is roasted and ground coffee.

## Equipment

Machinery for flexible packaging of portion-pack instant coffee is conventional horizontal form, fill and seal equipment used so widely for dry foods, e.g., Bartelt. Similarly, atmospheric pillow pouches of R & G coffee are produced on form, fill and seal equipment such as Mira Pak; FMC Corp., Stokeswrap; General Packaging; Hayssen; Package Machinery Corp., Transwrap; and Triangle. All form the package from roll stock pulled or driven over forming collars, sealed on the back and end heat sealed on the top and bottom during filling.

Vacuum packaging equipment for coffee is a development target for several machinery companies who have had active projects. This is an illustration of response by equipment engineers to a perceived need.

The Hesser bag-in-box unit has been successfully used in Europe. The Hesser Vacufin operates at 80 pm from roll stock aluminum foil laminate. After being filled the bags are evacuated in a separate chamber prior to sealing. The bag is in the carton during all operations, with carton closure taking place after bag sealing.

Container Corporation of America's CekaVac system accomplishes the same end but uses a knocked down preformed bag-in-box as a starting point. The combination is snapped open and a bottom seal is made. The product is filled and a vacuum drawn prior to sealing. Initial packaging material cost is, of course, higher than that of the Hesser system.

FMC Corporation's Stokevac system forms a pillow package on a vertical form, fill and seal machine. The pouch is transferred to a vacuum chamber where evacuation and sealing occur. No carton exterior is used in this system.

The seemingly ubiquitous SIG has designed a SIG-O-MATIC system which forms packages from roll stock.

Mira Pak has developed a system which is a MiraMatic, redesigned for gas flush packaging.

Pneumatic Scale Co.'s imported vertical form, fill and seal Rovema equipment can produce 27-30 2-lb pouches pm and 60 1-lb pouches pm, the latter with a residual oxygen content of under 2%. As with most of the other machines, filling is by auger feed.[12]

The Ultra-Pak tetrahedral packaging operation of Hayssen developed an R & G coffee packaging application for their equipment. The tetrahedral package saves a significant amount of packaging material and, consequently, cost. A long forming section also minimizes stress crack damage to the material by eliminating the severe folding over forming collars.

These are illustrations of the amount of effort taking place in the area of flexible vacuum packaging of roasted and ground coffee. Other companies are doubtless active, some with proprietary developments. The quantity and quality of effort does not guarantee profitable returns. There is no question, however, that the concept will be given a thorough technical and economic testing in the real world.

# CONFECTIONERY

**Summary**

Candy products include box' chocolates, molded chocolate bars, enrobed bars, sugar candies and numerous specialties. They are manufactured by numerous old-line companies, few of which specialize and many of which make all types of candy products.

More retail outlets exist for candy than for perhaps any other food product. As most candy companies are small, the distribution system is complex and out of the hands of the candy manufacturer.

Candy products are, in general, relatively stable. Chocolate products require protection from fat seepage through the package. Soft sugar candies need protection against moisture entry or loss. High boiled sugar candies should be protected from moisture gain, and mint flavors must be protected from transferring off-flavors to other products.

Box chocolates are packed in fluted glassine cups with glassine and paperboard packing pieces in paperboard cartons. There is increasing use of thermoformed multicavity PVC trays for box chocolate packing.

Bar goods are generally packaged in glassine, two-piece if molded chocolate, one-piece if enrobed. Interior packing pieces are sometimes used for added greaseproofness.

Sugar candies, which require moisture protection, are dump filled in wax paper lined paperboard cartons or wax sealed in roll-style packages. Tabletted dextrose pieces are in sealed aluminum foil envelope pouches. Much sugar candy is individually wrapped in cellophane, polypropylene or waxed paper.

Bite-size candy is often in vertical form, fill and seal packaging, formed from pouch paper, duplex cellophane or polyethylene, depending on the contents.

Growth has been experienced in many sectors of the candy industry, but packaging innovation has not been a strong point of the industry.

**Introduction**

The last American food industry to maintain a fixed retail price for its mainstay product line was candy. This fixation on five cents as a figure never to be violated led to some of the most intensive pressures for truly low cost packaging materials the country's food industries ever have or ever will witness. It also led to the odd phenomenon, in the decades of continually rising prices, of candy bar package sizes increasing and shrinking regularly in response to fluctuations in cocoa bean prices. With a break in the pattern beginning to emerge, the confectionery industry may stabilize into more market oriented products and packages.

Tradition is a candy industry characteristic so deep as to defy the plans of some of the larger corporations who have bought into the industry to diversify and take advantage of the profits. Where new marketing techniques have been employed, successful returns have been realized. But new marketing techniques are alien to many and unaffordable to most who participate in the industry.

Although other food industries can boast of derivation of sales and profit volume from products introduced in the past five or ten years, the candy industry can point to no such record. In fact, almost all sales and profits stem from products which were successful prior to and during World War II.

The candy industry can be divided into two large segments: box chocolates and bar goods. In the former category are the high and low priced individual pieces, usually chocolate enrobed, carefully nested in cups or compartments and boxed. In the latter category are the bars, sugar candies in rolls and lined paperboard cartons, and the chocolate pieces dump filled in windowed paperboard cartons. This category also includes a variety of jellies and hard candies in plastic bags.

Candy industry products might be divided into molded chocolate bars; chocolate enrobed pieces, including bar; bite-size centers enrobed in confectioners' coatings; hard sugar candies, including pulled and pressed; and soft sugar candies such as caramels, jellies and gums. Often included in the confectionery business statistics is chewing gum, both stick and candy coated. Edible nuts (not for cooking) are also often included, although some consider these snacks.

The products are retailed in a broad range of prices from a penny apiece to four dollars or more for a pound box. The major parts of the business, however, are centered in the five-, now becoming ten-cent bar goods, the $0.29-0.49 dump filled supermarket pack, and the $1.50-$3.00 a pound box chocolates.

The candy industry can boast the largest retail distribution system of any food industry in the United States. Candy appears in supermarkets in two locations, in its own section and at the checkout counter. Candy is sold in candy stores, drug stores, newsstands, from vending machines and so forth. From this quantity and diversity of retail outlets, one may infer the wholesale distribution network is complex. This is true, especially in the light of the great number of candy manufacturers who are too small to support their own sales forces. A system of brokers, wholesalers, and direct distribution is operative.

The existence of different classes of trade dictates a diversity of packages and packaging. The cost of critical ingredients, such as chocolate or sugar, has a key influence on the weight and, consequently, on the size of the bar. Changes in size, of course, mean changes in packaging for the many different packs involved.

In box chocolates with (formerly) fixed retail prices and with fixed size set-up as containers, the product mix is changed to respond to ingredients cost changes. This leads to changes in internal packing materials.

Labor costs have been steadily increasing. These significant cost rises have been interrupted by the installation of mechanization and even some automation in candy plants. Still, a considerable amount of hand labor continues to be used for packaging candy, particularly in box chocolate packing. The problem is a combination of lack of capital to invest in new equipment and lack of suitable equipment.

Even the larger manufacturers often have heavy labor costs, partly due to a lack of automatic equipment and partly because of a broad range of products. Many candymakers manufacture a host of products to satisfy the demands of their various customers, and the volume of some product lines is so small that no machine is justifiable.

Candymaking is a relatively easy manufacturing process, as is amply demonstrated by the numerous one-man retail candy manufacturing operations throughout the country. Large inventories of used equipment are available for use in adding new products or in starting new candy businesses. Brokers are willing to add products to their lines.

Because of the ease of entry into the business, there is fierce competition in the secondary volume areas and this is eroding the position of the leaders. There is no dominant company or group of companies in the business. Some are large, but hold no consumer franchise. Some hold a consumer franchise but are not large by food industry standards. There is generally an obvious leader in each product and market category: box chocolates, enrobed bars, sugar candies, chewing gum, nut rolls, coconut, etc. Leadership in one category does not mean leadership in any other category, although in a few instances a company can boast leadership in more than one product group. Into this maze, directly competitive, secondary branded products and, of course, private label products have become significant factors. Price products are common in all classes of market and product. Local and regional producers abound, although the national brands tend to dominate in each product class; a few regional brands are, however, quite strong.

Most of the smaller candy companies make, package and sell candy exclusively. In larger companies, chocolate making for industrial purposes may be part of the operation. Candy operations which are part of large organizations generally function as independent entities within those corporations. There are few candymakers who have other products in their manufacturing or distribution systems.

In some instances, packaging is imitated to take advantage of packaging material production runs, as well as brand identification. At one end of the candy packaging spectrum is inexpensive packaging for low cost products. At the other end is elaborate, expensive packaging used for special event and gift candies such as Valentine's Day or Mother's Day candies.

The edible nut marketers are in a multiplicity of businesses ranging from commodity trading to peanut butter to nuts for other food producers. Roasted peanuts, the leading product, is dominated by one major company selling both supermarket size and smaller five- and ten-cent sizes through candy channels. Almonds, the second most important product, are controlled by a growers cooperative, but are marketed through the several nut roasters. Other nuts include pecans, cashews imported from India, and filberts. A few nut roasters are national, but have competition from a large number of local and regional operators.

## Packaging Requirements

The nature of the candy business makes one fact very clear: the most important packaging requirement is cost. Other cost variables, such as ingredients, labor, and distribution, are somewhat beyond candymaker control; packaging is an area on which pressure has been exerted for cost reduction. Very small weights, such as fractional ounces, are individually packaged. Thus, packaging materials are carefully examined for their economics.

Fortunately, most candy products in their marketable state do not support microbiological growth. Thus, refrigeration, freezing, and hermetic sealing to avoid microbial contamination are not necessary.

Further, most chocolate and chocolate-flavored confectioners' fat coated products are rather well protected from moisture uptake. Solid chocolate and enrobed goods do not require high moisture impermeability in packaging materials or constructions. Chocolate and its analogues can, however, lose moisture and may be directly damaged by moisture loss over longer periods of time and under extreme situations. Coatings are not always intact and can, therefore, pass moisture through voids. White sugar bloom can develop on the surface of chocolate which has been exposed to high moisture.

Centers are made to retain the amount of moisture originally present. Loss of water from a soft, chewy nougat leads to hardening; from a cream to non-conversion to the desired fluid center; from a caramel to brittleness. Gain of moisture by a nut center removes crispness, by a brittle to chewiness, by a honeycomb to such extremes as fluid center.

The moisture problem with chocolate items is relatively minor compared to the fat problem. Milk chocolate, the most prevalent in the United States, contains low-melting butterfat in addition to cocoa butter. Handling practices can be abusive. Temperature elevation leads to fat melting which, in turn, can be visible on the package if there is no barrier. Fluctuations in temperature can lead to fat bloom on the surface, a defect which cannot be prevented by packaging, but which can be hidden by opaque packaging. Greaseproofness, or at least opacity of packaging material, is therefore a prime technical requirement of chocolate.

Cocoa butter and butterfat are relatively oxidation stable, so oxygen barriers are not warranted. Both cocoa butter and butterfat, however, are very good odor absorbers and require an odor barrier to minimize the danger of off-flavor pickup.

Chocolate's desirability is due partly to its surface gloss or appearance. Because chocolate is soft, it is readily scuffed, scratched, chipped and even broken. Packaging material should be sufficiently smooth and soft to minimize surface damage. Structural packaging pieces are needed to minimize crushing, denting and breaking.

Because chocolates and chocolate enrobed bars are sold individually and displayed at the retail level, surface graphics are a necessary part of the package.

Sugar candies present a totally different problem. Pectin and starch jellies can lose moisture which contributes to their softness. Most jellies are sanded with sugar crystals which are abrasive to packaging materials.

Hard sugar candies are non-crystalline, hygroscopic sugar and corn syrup glasses. Moisture absorption leads to graining and chewiness of the hard, brittle product. Fluid sugar and corn syrup can become very sticky and adhere tenaciously to the interior of packaging materials. Moisture exclusion is, therefore, a critical requirement of hard sugar candies.

Many pressed sugar candies have strong mint flavors which are easily transmitted to and absorbed by other products. Similarly, high-boiled, hard, sugar candies contain flavors which can be transferred to other products or which, in extremes, can pick up off-flavors. Tabletted hard candies are somewhat fragile and require some sort of structural support or regular alignment to assure against rattling. High-boiled hard candies are rather strong pieces which can withstand considerable impact unless they have been formed into intricate shapes.

Caramels and fudges contain moisture which contributes to their soft texture and chewiness.

Gain of moisture can lead to fluidity and stickiness. Loss of moisture can lead to crystallization and hardening and, in the case of caramels, very difficult chewiness. Individual pieces are wrapped because they would otherwise stick to each other.

Chewing gum is subject to moisture loss with resulting brittleness and so it requires a high degree of moisture protection. Again, pieces are individually wrapped. Sugar coatings on chewing gum are relatively stable but can lose gloss and crispness from moisture pickup.

Most nuts used in candies and eating nuts are roasted in oil for flavor and texture. Moisture gain leads to loss of desirable crispness and flavor. Moist nuts are also susceptible to accelerated rancidity of both nutmeats and the oil picked up during roasting. The changes are relatively slow, so, although desirable, protection against atmospheric oxygen is not often built into nut packaging. Nut packages, however, must be greaseproof to prevent passage of roasting and nut oils.

Another class of candy-like snacks is caramel-coated popcorn. Caramel can readily absorb moisture and become sticky and cause sticking together of individual pieces. Moisture protection is thus a requisite of packaging sugar coated popcorn and nut products.

**Packaging—Box Chocolates**

Box chocolate packaging falls into two broad categories: in-store and factory. Only the latter is a subject of this discussion. Factory packaging includes two styles of paperboard carton: set-up or corner-stay and folding cartons formed automatically or manually from die-cut flat blanks with lock or glued corners.

The candies, generally chocolate enrobed pieces and a few molded pieces wrapped in aluminum foil (for decoration and for keeping fluid centers from leaking), are place packed within the box. Two methods of ordering are in use today: the traditional placement usually by hand (but in a few plants by a complicated machine) in fluted brown glassine cups individually and placement in a multicompartmented, thermoformed plastic tray. Glassine cups have little vertical strength, so paperboard packing pieces are required to prevent crushing in multitiered boxes. Further, glassine-faced cushioning is used to reduce scuffing of exposed tops of

chocolate pieces. Glassine cups, used because of appearance and greaseproofness, may be purchased nested or candymaker-formed mechanically from layered glassine sheets. A slow, but significant, increase in the use of thermoformed PVC and XT polymer trays is affecting the growth of glassine cups.

The printed, and often embossed, paperboard cartons are generally overwrapped in a transparent film. The overwrap serves to protect the candy from dirt and dust (box chocolates are made many months ahead of sale), to hold the top and bottom of a two-piece carton together, to prevent pilferage, to protect the carton decoration from dust, to brighten the carton graphics and to provide a medium on which seasonal messages can be easily printed. Cellophane overwraps were formerly the only type used and are still used because of their printability. Shrink films, such as polypropylene, are frequently used because of the permanent tightness afforded. Cellophane bands are used for promotional decoration.

**Packaging—Molded Chocolate Bars**

The bar goods category which contains the largest selling individual items is molded or solid chocolate bars. These products are manufactured with and without inclusions such as almonds, peanuts, crisped rice, and raisins. Solid chocolate bars are made from tempered fluid milk chocolate. The fluid, with or without inclusions, is deposited in molds in which the chocolate is solidified by cooling.

The classical milk chocolate bar wrap is two-piece—an inner white opaque glassine in a die-fold form and an outer printed paper, usually a sleeve, but sometimes an overwrap. The glassine is a greaseproof barrier which peels away easily from partly melted chocolate, a not uncommon consumer problem. The paper sleeve holds the glassine together, serves as a second barrier and is a good printing substrate. With a two-piece wrap, the inner glassine need not be sealed together, and so the bottom appearance of the chocolate is not damaged by the packaging operation.

Perhaps the major reason for using the two-piece wrap, however, is tradition. It was used many years ago by the leader and continues to be used despite some changes.

Several variations on the most common

package are in commercial use. Miniatures use unmounted, aluminum foil dead fold wrapped against the chocolate. The pieces are so small that die-fold wrapping of glassine would be difficult. Aluminum foil is a satisfactory grease barrier although it is somewhat difficult to remove from chocolate which has partly melted or softened. A printed paper sleeve is placed over the foil.

One manufacturer of a molded shell piece uses glassine wax laminated to aluminum foil as the inner sheet with a printed outer litho paper sheet. The glassine in contact with the chocolate is the greaseproof barrier; it serves as a good release from the chocolate, especially when laminated to aluminum foil. The wax laminant serves as a smooth bonding agent to provide good machinability. It further keeps the plies together sufficiently to exclude fat which causes delamination with other bonding agents. The aluminum foil exterior provides some dead fold, some moisture protection and considerable decorative effect.

The popular American chocolate kisses or conical shaped, solid chocolate pieces are deposited in that shape and solidified by chilling. They are subsequently dead fold wrapped in unmounted aluminum foil with a paper pull tab opener. The foil acts as a peelable fat barrier and, of course, as a highly decorative wrap. Foil-wrapped pieces are dump filled into vertical form, fill and seal printed duplex MS cellophane pouches. Little moisture protection is needed, but the foil wrapping is evidently believed to be a marketing advantage. Transparency of the overwrap is thus warranted. Single-ply cellophane could be used, but any tear from either foil scratching or pilferage attempts would be easily propagated; duplexing multiplies the strength without significantly decreasing the transparency.

**Packaging—Enrobed Bars**

Because they are better consumer values in terms of weight of candy per cent paid, enrobed bars enjoy the greatest sales volume of any confectionery class. These bars consist of a preformed center of nougat, fudge, nuts, caramel, cream or some combination. The relatively stiff center is covered with either chocolate or an analogue, such as confectioners' fat (cocoa butter is replaced by another vegetable fat). Cen-

ters are composed of less expensive ingredients than are coatings. Therefore the total ingredient cost for a given weight of product is lower than for solid chocolate. An enrobed bar is also less difficult to manufacture than a solid bar.

From a packaging standpoint, enrobed bars are irregular in size and shape (and weight). They generally have flat bottoms and rippled or decorated tops. (Smooth uninterrupted chocolate is somewhat dull unless it has been molded.) Products such as nut rolls are totally irregular because the centers have been rolled in nuts prior to coating.

Enrobed bars are usually packaged in the least expensive greaseproof material available, which is glassine. Most are packaged in singleply, bleached or opaque, surface printed glassine using either a die-fold closure or a pouch style with the end fins folded under. Closure is by adhesive or by thermal spot on the base of the die fold—all in line with the important cost considerations of a five- to ten-cent retail selling price on one-half to one and one-half ounces of product.

Many candy bar manufacturers use a piece of die-cut chocolate paperboard as a base beneath the bar in wrapping. This piece acts as a stiffener and a grease barrier as well as a material which hides the fat penetration, and provides structural support in tiering. It is apparently a somewhat unnecessary packaging piece for the rectangular solid bars, but serves a very useful purpose with round and odd shaped pieces such as turtles or peanut butter cups. The card provides a regular shaped base as well as a regular size for wrapping otherwise totally irregular pieces.

Several candy bar makers use a second interior sheet of greaseproof or other paper between the bar and glassine. This makes the glassine appear opaque. Otherwise, the glassine, in contact with the darker chocolate, often tends to appear oily.

The glassine is surface printed by flexographic or rotogravure process and generally overlacquered for gloss and protection of the printing. Single-ply glassine acts as an inexpensive, machinable, greaseproof, decorative material which satisfactorily protects candy bars from dirt. The material is only barely moistureproof, and the closure hardly sufficient to exclude odor or moisture. However, glassine is an effective package in that it fulfills the objective of providing point-of-purchase merchandising.

A number of enrobed salted nut rolls are wrapped in printed transparent film to display the desirable appearance of the nuts. Other products in this class require release-coated interior paper sheets because the nuts are adhered by caramel which can be extremely sticky when not protected against moisture. Both cellophane and silicone coated papers act to minimize sticking.

Peanut butter cups are molded directly in fluted glassine cups. The cup is wrapped on a chocolate board in glassine in a bunch wrap. The glassine inner cup acts as an easily strippable mold and provides greaseproofness. The bottom board is present to allow for stacking without nesting and attendant tearing of the wrap or marring of the chocolate surface.

Bars, whether molded or enrobed, cups and salted nut rolls are often packaged in various sizes in multiples. The most common is the six-pack, a neat appearing pack designed for supermarket shelf display. Bars are neatly aligned in a paperboard tray in one tier of six or two of three. The filled tray is overwrapped with printed or unprinted transparent film to display the several bars. Cellophane is used most often as the overwrap, although biaxially oriented polypropylene may be used. With biaxially oriented polypropylene, however, the seals are not as neat as with cellophane.

Ten-packs are designed to have the appearance of economy and so employ packaging which appears inexpensive. Wrapped bars are dump filled into surface printed polyethylene bags which may be sealed or tie-top closed.

Some multipacks have somewhat more elegance (and bag durability) built in by affixing a paper header to the closure. The combination of bars adjacent to each other and manual header application tightens the package and decreases the movement of bars.

Miniature or small-size enrobed bars started with the Halloween demand and are still dominant during the Halloween season. Enrobed miniature bars are packaged in glassine as are full-size bars. Although five- and ten-cent bars are packaged in 25-30 lb per ream glassine, smaller bars have sharper folds and creases and so require heavier weight glassine, in the range of 35 lb per ream. Pouch-style packaging is often used because of the small size. Miniatures are dump filled in vertical form, fill and seal or in preformed surface printed polyethylene bags. Again, cost is an important consideration and polyethylene provides the needed containment inexpensively.

Numerous variations on these packages are in use and will continue to be. Flexible packaging materials and constructions provide the required protection at low cost. As the retail price of candy increases, some upgrading of candy packaging might occur, but the change will be slow due to the historical adherence to tradition.

### Packaging—Sugar Candies

As indicated above, the problem with sugar candies is one of protection against moisture transfer. With the mint flavors, aroma transfer is a factor to be reckoned with. Sugar candies are low ingredient cost products which generally cannot be eaten at one sitting. Reclosability of the package is another parameter.

Jellies are packaged in much the same manner as enrobed bars. The overwrap is surface printed, transparent cellophane, used because of its moistureproof nature and to display the colorful pieces. This is not an outstandingly good moisture barrier, and, as a result, significant differences in shelf quality are frequently observed.

Flavored gums are packaged in a unit closely resembling the double package used for cookies and cereals. The size is, of course, considerably smaller. Candies are dump filled into waxed paper lined, tuck-end, paperboard sleeve cartons. Fold closure of the interior liner provides fairly good moisture protection. Cough drop versions of gum candies have MS cellophane overwraps; these command premium prices.

Similar packaging is employed for a variety of hard sugar candies which are dump filled. Hard sugar cough drops are packaged in this manner but with the MS cellophane overwrap for added protection. A cellophane tear tab is usually an integral part of the overwrap. The carton plus liner affords reclosability for the sugar candies which are consumed slowly over long periods of time.

The most widely known package for sugar candies is the roll pack. Disc or doughnut shaped candies are aligned face to back in the form of a long cylinder. The cylinder is wrapped in a waxed paper aluminum foil combination, heat sealed along the length and crimped at the ends. Heat applied to the ends causes wax flow to seal the ends against moisture. A tear tab is usually employed to ease opening. A printed paper sleeve is employed for graphic identification. Foil on the exterior provides some moisture protection, and a great deal of dead fold to hold the ends in place before, during and after the wax flow. The wax appears to be the major source of moisture protection, particularly by filling the voids in the ends. Paper helps protect the foil as a moisture barrier.

Many sugar candies are individually wrapped when they are dump filled in pouches or bags. This helps provide moisture protection. But more important, this prevents the pieces from sticking to each other in the event of moisture entry. Wrapping may be twisting in which the material is folded around the piece and the ends twisted shut. Mechanically, this operation requires material with good dead fold characteristics. Aluminum foil is, of course, excellent but expensive. Some premium-priced, high-boiled sweets, however, use aluminum foil combinations with waxed paper. Cellophane was the material of choice until a few years ago because of its excellent high speed machinability (the equipment operates at over 600 pm) and good dead fold.

Cast polypropylene was found to have better dead fold characteristics for twist wrapping, and better moisture barrier properties. Further, the material is cheaper than cellophane, and thinner gages may be used to allow for improved machine efficiency.

Alternatively to twisting, individual pieces may be wrapped in a caramel wrap. Pieces wrapped in this manner are generally square or rectangular. Round, oval or irregular, as well as rectangular shapes employ twist wraps. Caramel wraps are, as much as possible, heat tacked at the bottom to keep the ears down. For this reason, materials such as MS cellophane or waxed paper are used.

At least one major hard candy maker uses caramel wrap for the square inner pieces. The pieces are aligned in a rectangular block which is overwrapped in a printed aluminum foil paper/wax/strike-through tissue construction.

Lacquer on the foil surface is compatible with the wax which acts as a moisture protection; foil's dead fold characteristics provide reclosability.

Many wrapped hard sugar candies, being relatively inexpensive products, are dump-filled in inexpensive pouches formed from cellophane or polyethylene. The cellophane is usually duplexed for added strength. Duplexing also offers the opportunity for reverse printing of the outer sheet of the pouch.

Polyethylene bags for hard candy packaging need only be single-ply because of the strength of this plastic film. Because of a pilferage problem at the retail level, candy bags are usually heat sealed rather than tie closed.

Lollipops are wrapped in two different manners. These are generally hard sugar candies with a handle affixed for eating convenience. At least one manufacturer twists a printed waxed paper over the candy portion allowing the stick to protrude. The more common method is to heat seal two cellophane sheets around the candy perimeter, allowing the stick to come through a small opening. The MS or polymer-coated cellophane provides moisture and aroma protection, and release properties in the event of moisture gain. Wrapped pops may be dump filled into polyethylene bags. Extensible plastic is needed because of the stick's ability to penetrate a more brittle film such as cellophane.

Unenrobed caramels, being sticky, require a moistureproof release material. Caramels, whether milk or chewy sugar pieces, are individually wrapped. Cellophane is generally used for milk caramels. Wrapped caramels may be dump filled in film pouches. A frequently used pouch is made from cellophane/polyethylene laminate to provide the required moisture protection.

Wrapped caramels are packaged in sticks for five- and ten-cent sale. The caramels, butted against each other, are cellophane overwrapped using printed, heat sealing cellophane. The outer overwrap acts as the principal moisture barrier. The inner wrap prevents sticking of pieces to each other.

The abrasive action of sugar sanded jellies on packaging interiors has restricted the packaging forms used. Recently one company has employed a coextruded film on form, fill and seal equipment to provide the needed combination of moisture and abrasion resistance.

Tabletted dextrose candies have high moisture protection requirements because of their susceptibility to deterioration. The tablets are packaged in paper/aluminum foil/polyethylene envelopes to provide the requisite moisture protection and reclosability.

### Packaging—Specialties

A number of confections which do not clearly fall into any of the above categories are packaged for the American market. Moisture protection for licorice is imparted by cellophane pouches or, in the case of small individual pieces, by MS coated pouch paper in pouch form.

Bite-size sugar coated chocolate, as stated above, does not require high moisture or fat impermeability. The pieces, however, are hard, dense, and heavy and can not be adequately retained by a glassine pouch. For this reason, small pouches are formed from printed MS coated pouch paper which provides some moistureproofness, heat sealability, and the strength to retain the product. For economic reasons, larger size pouches are duplexed MS cellophane or duplexed MS cellophane and MS pouch paper. The cellophane is reverse printed to obtain a better graphic effect.

Caramel-coated popcorn and nuts appear in two highly moistureproof package types. The older type employs waxed chipboard overwrapped with aluminum foil/paper/wax/strike-through tissue in which wax flow is employed to effect a moisture seal. A newer type is the bag-in-box construction, using a paper/aluminum foil/polyethylene preformed inner bag.

### Packaging—Chewing Gum

Packages containing more than one sugar-coated piece are packaged in flat paperboard cartons with cellophane overwrap. The flat carton keeps the pieces aligned to prevent chipping.

Each individual piece of stick chewing gum is wrapped in an aluminum foil/paper lamination which keeps the sticks from adhering to each other. There is no tacking or sealing on this package.

Five or more sticks are wrapped in one of the finest overwrap systems in American packaging. The lamination employed has aluminum foil on the interior topped by paper which in turn is protected by cellophane. A conventional double point end fold overwrap scheme is used, but wax flow closes the long and end seals. The resulting package is as moistureproof as the barrier material allows, with no voids arising from the packaging operation.

### Packaging—Nuts

Most edible nuts are vacuum packaged in cans after oil roasting and salting. Dry roasted nuts are vacuum packaged in glass jars. Cooking nuts are packaged similarly. Many cooking nuts, such as walnuts, are anti-oxidant treated and are packaged in form, fill and seal duplex cellophane pouches.

Roasted nuts for small retail packages are packaged in a variety of package types which are inconsistent with the requirements. Most are vertical form, fill and seal pouches made from printed MS cellophane. The cellophane acts as a moisture and grease barrier, but not as a good gas barrier, which is required. Much of this product is rack-jobbed and, thus, is under the roaster's control. But this still does not account for a substantial proportion which is sold through candy jobbers.

More recently, dry roasted nuts are being packaged in sophisticated flexible vacuum envelope pouches constructed from film/aluminum foil/polyethylene. A small amount is in tetrahedral packages of paper/aluminum foil/polyethylene.

### Equipment

Box chocolate packing has been the subject of several isolated equipment developments over the years. The sight of a line of dozens of women removing pieces from bulk packers, inserting each into a glassine cup and subsequently placing them in paperboard cartons has moved inventors and engineers to attempt to mechanize the operation. The variation in size and shape of the many pieces has been one limiting factor. Chocolate pieces are relatively delicate both in structural strength and in surface characteristics. They cannot be easily moved without potential cracking, crushing, denting or scuffing. Were each the same size and shape, mechanization would be quite feasible. Were each individually wrapped, as some are in foil, their movement would be feasible. Box chocolates, however, have been marketed

51

# FIGURE 2

## Properties and Approximate Costs of Typical Materials Employed in Flexible Packaging Structures

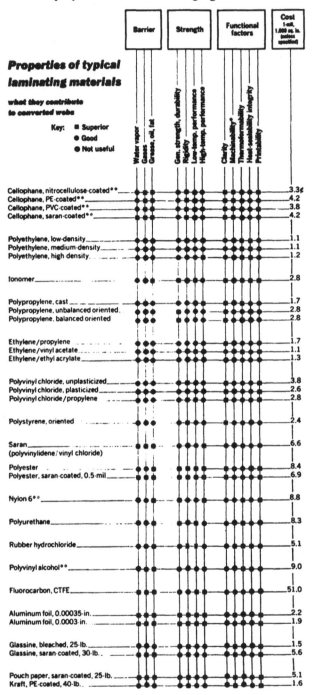

*Machinability generally is a function of strength, rigidity, slip, low stretch, low static.

**Barrier may be highly dependent on relative humidity; values shown for 50% RH or under.

X Depends on plasticizer content.

This tabulation is compiled from industry sources. For detailed data, see the MODERN PACKAGING Encyclopedia Issue, September 1967, pp. 102, 148, 181, 194, 197 and 203, or consult supplier data sheets.

From *Modern Packaging*, Oct., Nov., and Dec., 1967. With permission.

as variety packs (even to the point of not identifying the pieces). The problems of box chocolate packing are dramatically illustrated by the presence of fingerprints of packers on many pieces.

Among the several mechanisms developed for this purpose are foam padded vacuum-pickups and drops. This system, made by Kamen, relies on a fixed position for the box and a fixed cavity in the box. To achieve this positive location, thermoformed PVC trays are employed. The complexity of the Kamen system may limit it to larger box chocolate houses. Its installation reduces the employment of glassine cups and packing pieces in the future.

One major box chocolate manufacturer employs only one installation developed for pieces in glassine cups. Indications are that this type of system will not be installed in other box chocolate houses in the foreseeable future.

Chocolate and enrobed bars are packaged on a limited range of medium to high speed machines. Much of the equipment now in operation is old and very reliable, i.e., not likely to be replaced until it is totally obsolete or wears out. This situation has been a problem to the machinery manufacturers who developed and built such durable equipment.

Package Machinery Corp. wrapping machinery is well-known and widely used by bar manufacturers. An increasing amount of imported SIG equipment is being installed for two-piece molded chocolate bar wrapping. Speeds of up to 140 pm are nominal (and actual) for these intermittent motion machines.

Several new candy plants have installed Lynch equipment which forms die-fold packages. Lynch machines can operate with or without chocolate board and with or without adhesive. Lynch equipment has been designed to be integrated with six-pack or 24-count paperboard cartons. Bar wrapping speeds of over 100 pm are nominal for this equipment.

Improved SAPAL equipment is able to form a totally sealed package with fin seals.

The venerable Hudson Sharp Campbell wrapper, which makes pouch-style wraps, has been a mainstay in many enrobed bar plants.

This continuous motion machine wraps bars with either overlap or fin seals, and with or without end seal tucking. Speeds of 60-120 pm are nominal for this equipment which has been particularly useful for miniature bars for many companies.

Newer versions of SIG; imported Package Machinery Corp., Forgrove Rose; and Doughboy machines form the same style packages as Campbell wrappers.

Sugar candy individual wrapping is almost all performed on foreign equipment, such as the Rose twist wrapper from England, or the Latini pop wrapper. One American company manufactures a caramel wrapper, namely, the Ideal Wrapping Machine Company.

Packaging of wrapped (or unwrapped) sugar candies has been an important area of concentration of Crompton & Knowles Package Machinery Division, which makes equipment to form, fill and close wax paper lined tuck end paper board cartons, and to roll package. The C&K units have integral filling devices.

Vertical form, fill and seal equipment used by the candy industry is much like that used by other food industries: Transwrap, Stokeswrap, MiraPak, Hayssen and others. Filling is volumetric or gravimetric with the former dominating because of product value considerations.

The newer tabletted dextrose candies are packaged on modified horizontal Bartelt form, fill and seal equipment.

As in many other characteristics, candy industry packaging equipment is a mixture. European equipment makers have designed and built numerous highly sophisticated and specialized packaging machines for the confectionery industry. American manufacturers, for the most part, looking at a broader total packaging business, have made machines versatile. The exceptions in the United States have found their markets limited by the number of potential customers.

Several concentrated development efforts have been made by some equipment companies; these have brought sales returns. The innovative packaging which has resulted, however, has been quite limited.

# CONVENIENCE FOODS

## Summary

The term convenience foods is applied to a broad product mix and includes both unit packed foods for HRI use and dry foods for rapid consumer reconstitution.

Portion packaging, although originally rigid cup plus flexible peel seal, is turning to flexible pouches because of lower cost. Dry, stable products are in paper/polyethylene envelope pouches packed on very high speed horizontal equipment on a custom basis. Fluid flavorful condiments are in similar type packages with added film barrier but, because of lower quantities, are packed on vertical equipment. Increases in use of flexible unit packages for both HRI and consumer use are forecast because of labor savings and convenience.

Consumer packages of dry foods, to which a liquid or solid is added to prepare a meal or a food, are most often in large envelope pouches in retail paperboard cartons. Paper/aluminum foil/polyethylene is a commonly employed lamination fabricated on horizontal form, fill and seal equipment. A decided trend towards this package type, with additional consumer-use containers attached, is readily projectable.

## Introduction

This classification encompasses a broad range of products marketed in varying degrees of preparation to consumers and to HRI operations. The term convenience apparently has favorable connotations to consumers and food service operators. It implies that some aspect of labor or difficulty has been removed from food preparation. Convenience for the consumer means factory preparation and packaging that is easy to use.

Portion packaging for the food service industry is a typical example of the fast growing and broad nature of convenience packaging. The consumer is familiar with portion or unit-size packages of condiments, such as sugar, salt, pepper, catsup, mustard, mayonnaise, jelly, and soy sauce. Each of these was formerly dispensed from bulk containers in public eating establishments. Now, the bowl of sugar and sugar cubes are rarities and envelope packages containing a teaspoon of granulated sugar are ubiquitous. Use of unit packages of other condiments is increasing rapidly; portion packages offer a degree of cleanliness not possible with bulk containers. Unit packages are also precise quantities that can be measured by cost conscious HRI operators and by consumers. Overestimates by either can be returned and used at a later time. They require less labor than bulk packages. Portion packaging, of course, is more costly than bulk packaging, but evidently this is out-weighed by the value of having portion packages.

The list of foods now being unit packed is growing substantially: tartar sauce, salad dressing, pickle chips, whole pickles, lemon juice, and synthetic sweetener. Each is packaged in a single serving size in a package that is easy to open, easy to empty and easy to throw away.

The food products themselves may be manufactured and packaged by conventional food processors, but a large amount is manufactured by food processors and packaged and/or distributed by contract packagers.

Regardless of the supplier, portion packages move through franchisors or food service wholesalers. Turnover in this distribution is relatively rapid as it would have to be to accommodate the needs of food service operations. This quick inventory movement is fortunate for portion packed liquid products which have relatively short shelf lives in these packages.

Portion packaging ranges from flexible constructions to thermo-formed in combination with flexible to metal plus flexible. Aside from sugar, the best-known today may be the thermoformed tub with peelable flexible lid.

Coffee cream and whiteners have a somewhat different pattern. The microbiological hazard posed by coffee cream in large containers has led to improved sanitary requirements and regulatory pressures to eliminate the bottle or jar. A variety of systems has been proposed, including refrigerated product filled in portion packs by dairies, sterile unit packs, and fabrication of non-dairy coffee whiteners. This is a case in which the position of portion packaging will greatly influence the future of flexible packaging materials. In-line thermoform, fill

54

and seal systems with and without asepsis, nested cups with and without aseptic filling, and tetrahedral packaging are all commercially employed at present.

Some HRI portion packages have reached the retail level: sugar packets are available both to consumers and food service operators.

The cascade of new products reaching the retailer in recent years has been dominated by a multiplicity of imitations. Among the innovative convenience products and packages have been milk shake mix, refrigerated cookie dough, toaster sandwiches, metered calorie meals and seasoned breadings.

Milk shake mix, also available in sterile liquid and refrigerated liquid form, is marketed dry with a rigid plastic shaker mug and tetrahedral packages containing the flavoring ingredients to which milk is added. The flexible packages are sold both with the mug and as refills.

Refrigerated cookie dough has enjoyed some success against the competition of cookie mixes and baked cookies. The prepared dough need only be cut into discs and oven baked.

A few shelf-stable toaster products resembling baked goods have spawned a family of products oriented towards specific market segments. These products consist of pastry exteriors and low moisture fillings, usually jam-like. The pastry may be flavored and may be shaped into animals for children or coffee rings for adults. The surface of the pastry may contain a heat-stable sugar icing for flavor and appearance. Thickness of the product is about one-half inch so that the product may be inserted in a toaster for quick heating.

The liquid versions of metered calorie meals have been in existence for over a decade and enjoyed a peak about five years ago. Metered calorie meals have now found a comfortable growth level in solid, liquid and dehydrated forms. They are aimed at weight watchers, persons who have limited meal times but wish to have a complete meal, and perhaps to other distinct market segments. Thus, this category contains liquid dietaries, canned precooked entrees, baked goods, and dehydrated products to which milk is added to prepare a full meal for consumption from a single glass.

Seasoned breading mixes were designed to allow the consumer to create the effect of deep-fat frying in the oven. Not only the method of application, but also the total food preparation operation in the home was made more convenient. The product is basically a mixture of breading, fat and spicing. Fish, chicken or veal is shaken in a bag containing the mixture.

## Packaging Requirements

With no consistency among products falling into the category of convenience foods, no specific packaging requirements can be attached to the category. The general rules of protection apply, but each product group must be treated individually. This is a particularly interesting group because it contains products which have not previously existed, so only guidelines can be established for packaging.

Further, and perhaps most important, the product concept definitely includes the convenience of easy access to the contents. In some instances, the package can be an integral part of the product (e.g., pressurized whipped topping). The combination of not having a prior history of protection requirements plus a need for packaging convenience affords opportunities to employ the most appropriate materials and constructions. Flexible structures are used not on the basis of past reputation but because they are best suited to the total requirements.

For cost reasons, HRI portion packages are generally flexible constructions. As indicated above, turnover is very rapid, so physical-chemical protection requirements are not as strict as they would be if longer shelf life were required.

Granulated sugar for portion packaging requires only protection against internal abrasion of the sugar crystals and a barrier against dirt and dust. Some moisture protection is desirable as sugar can absorb water, but evidently this is not a serious matter. The package must be durable enough to withstand jumble packing and handling. The material must be eminently machinable because the quantity of packages dictates extraordinarily high speeds. Ease of opening, of course, is needed.

Synthetic sweeteners for the HRI trade are either powder or tablet. In either case, small quantities are packaged. The hard, sharp cornered tablets are very small and packed one or two to a package. Powdered form, because it is intended for dissolution in coffee or tea, is somewhat hygroscopic.

Salt is used as a condiment sprinkled over the surface of a food. It is abrasive and crystalline, but usually contains additives which prevent caking in the event of moisture access. Thus, the major consideration for salt portion packaging is the ability to dispense the contents in a measured flow.

Pepper, the most popular spice condiment, presents a slightly different problem. A pepper serving portion is considerably smaller than salt. Further, pepper contains aromatics which are subject to volatilization and deterioration over long periods.

Jams and jellies were originally packaged in miniature rigid containers that allowed spreading on bread and crackers. Because jams and jellies are subject to mold growth, they are packed hot, and the packaging must be resistant to the temperature. Loss of moisture leads to product pastiness and even to sugar crystallization. Fruit flavors can be lost by volatilization.

Syrups have the same protection requirements as jams and jellies, but are more dilute, and thus flow more easily.

Catsup is a homogeneous, spiced tomato condiment. The high acid content retards most microbiological growth, but the product is still subject to this deterioration if exposed to airborne microorganisms. Flavor from spicing and the tomato base can be lost by volatilization or oxidation. The red color is susceptible to non-enzymatic oxidative browning in time. Portion packed catsup should, therefore, be protected against moisture loss, aroma loss and, as much as feasible, access by oxygen, although, again, the quick turnover minimizes this requirement.

Mustard contains even more volatile flavoring components than catsup, and a major constituent is oily. The odor barrier is important because of this condiment's potency. Similar considerations apply to other products such as mayonnaise, tartar sauce and salad dressing. Mayonnaise can also separate and oxidize.

Pickles present a different problem because they have a form and shape which should be preserved. The contents are highly acid.

Liquid coffee whiteners are subject to microbiological deterioration. Packages for these products should be compatible with the sterility required of conventional dairy packaging and should conform to public health regulations concerning packaging of dairy products. Refrigerated versions must be able to withstand both the low temperatures encountered and condensation of moisture on the surface. Dehydrated forms have high fat contents which can badly contaminate seal areas.

Dehydrated beverage mixes, such as milk shake and metered calorie meals, are marketed through conventional retail channels and so require long shelf life. Most contain such ingredients as dry milk solids, sugar, corn syrup solids, and flavoring. Moisture ingress would lead to caking, browning, and fat degradation.

Toaster sandwiches are adversely affected by either gain or loss of moisture. Further, the pastry exteriors have high shortening contents. Moisture passage should be inhibited in a barrier which is not affected by contact with fat.

Refrigerated cookie doughs for convenience of use are in cylindrical form to be cut into discs for baking. Their moisture and fat content are high and, if not refrigerated, can spoil due to microbiological activity or fat rancidity. Both moisture loss and oxygen access should be limited by the package. Further, the cylindrical form should be preserved.

The fat and spice contents of seasoned breadings make them susceptible to fat oxidation which can adversely affect flavor. Moisture ingress could lead to caking. The ancillary convenience bag must be strong enough to withstand vigorous shaking with a heavy irregular object such as a chicken leg.

The general packaging requirements for convenience foods, then, are ease of use, and protection of readily reconstituted products against moisture and oxygen. Because most are in individual portions, packages are small and use a large quantity of packaging material relative to volume of contents. Cost, therefore, becomes an important factor leading to a desire to employ flexible packaging.

## Packaging

Portion packaging's need for speed and machinability coupled with its comparatively loose packaging requirements has led to relatively unsophisticated flexible materials.

Sugar is in a polyethylene extrusion on white paper. The plastic provides reliable heat sealability, some moisture protection, and fair abrasion resistance. Many, if not most, sugar portion packs are imprinted with the designation of the HRI establishment as well as content identification. The number of such operations forces the packager to do his own printing rather than have the roll stock preprinted by a converter.

By far the greatest quantity of unit packed sugar is in three-side seal envelope pouches. Recently, straw-like tubes pinch sealed at each end have been produced by a contract packager.

Salt appears in a number of package styles, the simplest of which is a three- or four-side seal envelope pouch of paper-polyethylene extrusion. This package does not have metered dispensing as does the paper/fluted paper combination containing salt in a number of tubes created by the fluting. A transparent shallow thermoformed plastic is sealed to a flat paper backing in yet another salt unit pack. Both the fluted and thermoformed units require breaking of a tip or end to open a tube through which a limited amount of salt can be dispensed. Pepper packages are generally the same as salt packages but, of course, may contain smaller quantities of a more expensive product.

Jams, jellies and syrups, as indicated above, are usually in rigid vinyl or aluminum foil tubs; peel-off lids are flexible structure often containing aluminum foil. Printed paper is the common outer ply with a compatible heat sealable, peelable coating on the interior to provide a hermetic seal.

Catsup, mustard, soy sauce, lemon juice and similar flowing condiments are usually in flexible, four-side seal, envelope film pouches containing polyethylene on the interior. Reverse printed cellophane is a typical outer laminant. Some of these packages are one-side printed and others are transparent to display the contents. This is a somewhat risky procedure since the contents may be susceptible to light-catalyzed as well as other color changes. A limited number of packages employ an aluminum foil lamination with printed foil exterior, but the cost is higher, although shelf life would be longer. Some packagers use envelope pouches for sauces and salad dressings, but most are the vinyl cups described for jams and jellies.

Plastic and metal tubes, used successfully for fluid condiments in Europe, have not been commercially successful in the United States.

Liquid coffee creamers in unit packs are currently in a variety of packages. In-line form, fill and seal vinyl cups with paper/foil/sealant peelable lids are employed in a limited number of dairies. This high capital cost equipment may or may not use aseptic filling and sealing. Nested cup filling and sealing, with or without asepsis, is also used. Tetrahedral vertical form, fill and seal packages formed from two-side polyethylene extrusion coated, lightweight, SBS paperboard are being used by a number of dairies.

Non-dairy dry coffee whiteners are packaged in the same manner as dry condiments, i.e., three- or four-side seal envelope pouches fabricated from paper/polyethylene.[43]

Dehydrated beverage mix formulations, such as unit meal packages of metered calorie liquid meals, are packaged in conventional dehydrated food pouches. Three- and four-side seal, printed paper/aluminum foil/polyethylene provides good moisture protection and is easy to open. Paper on the exterior protects the aluminum foil from exterior abrasion and breakage. Paper also serves as an excellent printing medium and pouch stiffener for stand-up display. This laminate is quite machinable and has served as a good basic barrier for dry foods for many years.

One milk shake mix and a metered calorie mix use much the same lamination, but the package is in a vertical form, fill and seal tetrahedron with overlap seal. Large tetrahedral packages such as these must be specially arranged in order to reduce the volume occupied by these less than symmetrical flexible packages. Tetrahedrons in this size use about a third less material per volume of contents than a flat pouch. The usual marketing method for pouched beverage mixes is in multiples in a paperboard carton that acts as the principal point-of-purchase display media.

Seasoned breading mixes and toaster sandwiches use the same type of pouch as do other dry mixes: four-side seal aluminum foil lamination pouches. To reduce product damage from opening, a long tear strip is built into the toaster sandwich package.

Refrigerated cookie doughs have more stringent requirements than other food products in this classification. To achieve the desired cylinder devoid of oxygen, the dough is stuffed into a chub package on vertical form, fill and seal equipment. A tube of moisture and gas impermeable inelastic double-wound saran is sealed along a long overlap. The dough, under pressure, forms a cylinder with no headspace in order to exclude oxygen. Metal clip seals, top and bottom, complete the closure. Graphics is reverse printing on the outer saran film ply.

### Equipment

Because there were only a limited number of portion packagers in the United States, equipment of appropriate speed was not always available from suppliers. As a result, a large portion of today's portion packaging is performed on proprietary machinery of the packager's design and construction. This is especially true of thermoformed cup packaging of jams, jellies and syrups. A number of European machines for this purpose have been imported into this country and are being used at scattered locations. These include the Form Seal marketed by Anderson Bros. Mfg., Co.; Atlas-Vac; Hofliger & Karg; Sendform, marketed by Package Machinery Co., and several others.

Flexible packaging of unit packs, on the other hand, uses some known American technology and, although some proprietary equipment exists, is often performed on stock machinery. The Cloud horizontal form, fill and seal unit with a continuous synchronized gravity fill is typical for sugar and salt packaging. Bartelt equipment can also be utilized for this function. Speeds of 500 pm in a single line are nominal.

A commonly used machine for packaging fluids, such as catsup and mustard, is the Circle unit with multiple positive displacement fillers. This machine works from two rolls of material, one front and one back. A parallel series of face-to-face seal tubes is formed and filled. The tubes are cut apart vertically during filling and top-cross sealing, or may be perforated so that the packages may be handled in magazine or accordion fashion. Again, Bartelt offers equipment that can make similar packages.

Chub equipment for refrigerated dough packaging is manufactured by Kartridg Pak. It is vertical form, fill and seal with special back seal heaters and synchronized positive displacement feed to assure void-free filling.

Bartelt horizontal form, fill and seal equipment is by far the most widely used for fabricating envelope pouches from aluminum foil laminates. This, as well as its principal competition, Delamere and Williams sold by Pneumatic Scale Corp., is available in intermittent motion at speeds of 60-120 pm or continuous motion at up to 300 pm. An imported Höller machine preforms pouches and then transfers them to a filling turret. Bartelt and Pneumatic both supply a range of compatible fillers, which for dry products are auger or auger plus gravimetric.

Tetrahedral packages for dairy creamers are made on either Tetra Pak or Ultra Pak equipment; those for dry products employ Ultra Pak which apparently holds proprietary rights on the fin long seal.

# DAIRY PRODUCTS

### Summary

Dairy products include milk, ice cream, yoghurt and cheese. Many items have been sold as commodities for many years, but the number of items is expanding with new variations. The use of flexible packaging, except for cheese, is limited.

A new concept in milk packaging involves the use of polyethylene pouches made on vertical form, fill and seal equipment.

Ice cream novelties have moved out of preformed glassine bags into horizontal form, fill and seal pouch-style packages formed from polyethylene coated paper.

Cheeses include natural and process. Both are gas flush packaged in flexible materials to create an oxygen-free environment. Complex laminations including PVDC as the main fat, gas and moisture barrier are formed into pouch-style packages. Paper interleaving is used for natural cheese to reduce slice merger. Process cheese is cast into polymer coated cellophane or polypropylene in individual slices and then multipacked into gas flush vacuum packages. Obvious deficiencies still exist in cheese packaging.

### Introduction

Although penetrated in many segments by vegetable-based products, the dairy industry continues to be viable. For packaging and marketing purposes, there are few differences between cow milk products and the imitations.

The latter are nutritious and have most, if not almost all, of the desirable flavor and appearance characteristics of the natural counterpart. Technologists' ability to alter and build with imitations has led to new families of products not economically or legally possible with natural foods.

Fluid milk, butter, cheese and ice cream are still major products made and distributed by local and regional dairies under reduced temperatures. Many have elected to use the distribution channels to expand the product line to include juices and juice drinks, pressurized toppings and imitations. Larger regional and national manufacturers have developed and marketed processed and natural aged cheeses and cheese foods, fermented milks, non-fat dry milk and many other products derived from fluid milk.

Fluid milk has changed from a daily delivered, highly perishable commodity in a glass bottle to a commodity with a week's refrigerated shelf life due to improved dairy sanitation and equipment. Packaging now ranges from flexible through paperboard to glass in sizes from one-half pint to a gallon. Home delivery has declined in the face of supermarket and jug shop selling. Because milk is still a product in which microorganisms survive and grow, it is subject to spoilage, and is pasteurized and packaged near the point of sale. Studies continue on sterile and concentrated milks, but the volume/weight ratio of these products may limit distribution to ranges only slightly greater than those that exist today.

Butter, although it has slipped below margarine in per capita consumption, is made by a number of local, regional, and national dairies. Butter's low moisture content makes it far less susceptible to microorganisms than milk, so it can be shipped far greater distances. Refrigeration is used to maintain butter's form, and to minimize flavor deterioration.

Margarines are generally the province of vegetable oil producers. These companies use dairy techniques in manufacturing and distributing their product. More important, they are not bound by the traditions of the dairy industry and have employed new marketing techniques such as television advertising and packages that double as table serving dishes.

Ice cream blossomed with supermarkets and the increase in numbers of home refrigerator-freezers. Once the consumer had the ability to purchase and hold ice cream conveniently, a price competition started among all dairies for the consumer's ice cream dollar. The consumer benefited by getting larger volumes of ice cream at lower prices. Ice cream today is overwhelmingly a bulk commodity. Consumer mobility has created an increasing demand for ice cream and its analogues at the HRI level. Ice milk, frozen on premises, has enjoyed some healthy increases as a result.

Cheese has enjoyed a rebirth. Cottage cheese rode the weight watchers' fad, and was rediscovered as a food on its own merits. Cream cheese was perhaps found in the potato chip dip fad. Italian cheese rose with the rise of pizza and spaghetti after World War II. These cheeses plus integrated preservation packaging techniques helped to bring all cheese to a continually rising volume.

In spite of continuing research, cottage cheese is still a perishable product that must be manufactured near the point of sale. To improve its position, cottage cheese is now sold in different size packages, with vegetables added but, unfortunately, still as a commodity.

In contrast, lower water content, more shelf-stable, natural and processed cheeses are made and distributed by national manufacturers and many of these are prepackaged products. All cheese must be refrigerated to retain flavor, prevent separation of fat and reduce the possibility of microbiological deterioration.

The oxygen-free preservation techniques used for processed cheese and cheese foods (in which additives are combined) create foods which the consumer could not pry apart. Only recently has new convenience packaging been combined with gas flush methods to develop packaging from which product can be removed without too much difficulty.

The consumption of fermented products, such as sour cream and yoghurt, has also risen, with the former remaining a commodity used as topping. The latter, however, has been flavored, branded, touted for diet and even longevity, and is now being marketed frozen and packaged in serving dishes.

The real growth of non-fat dry milk for the retail market began with agglomeration processes which created a product readily soluble in water.

The dairy industry may be characterized by a multiplicity of local and regional fluid milk and ice cream producers who have perpetuated these items as commodities. Cottage cheese and sour cream, which are highly perishable, remain local dairy commodities distributed under refrigeration. National dairies have created products out of several natural and processed cheeses, all of which are also distributed at reduced temperatures.

### Packaging Requirements

Except for dry milk solids, most dairy products contain high proportions of water that are combined with protein, fat and sugar. Those with higher water contents, such as fluid milk, are excellent microbiological growth media. Furthermore, processing to destroy or remove naturally present microorganisms considerably degrades product quality. Pasteurization is a mild heat process to destroy potential disease organisms; heat sterilization has not been developed sufficiently to retain fresh milk flavor, chemical or physical characteristics. Thus, pasteurization is one of the few practical processes to reduce microbial counts. Natural biochemical alteration is employed in the creation of wholly new products such as cheese and fermented milks.

At the present state of knowledge, even when heat sterilization is effected, as it can be, changes due to storage apparently occur unless additives or refrigeration or both are used.

Fluid milk and its associated flavored, skim, low fat, filled and imitation milks are pasteurized by statute. Because all of these milks are liquid, pasteurization is performed outside of the package. Dairy equipment includes cooling areas, so fluid milk packages are not required to withstand high temperatures. Closure systems must be sufficient to prohibit microbiological contamination of contents. Mandatory today is the ability to open, dispense and reclose easily.

For reasons which are elusive, milk packaging speeds are relatively low, i.e., 60-100 pm maximum, and stacking and multipacking have not been considered in marketing.

Butter and margarine packaging must obviously be greaseproof and odor proof. Beyond this, butter packaging has remained very much the same for decades. When colored margarine was first legalized, packaging followed that of butter. Today, however, opening, use, reclosure, attractiveness, multipacking, and air-tight seals are becoming more important.

Ice cream is packed in a fluid form, so the package must be liquid tight. The package is then closed and hard frozen; thus, it must be able to withstand temperatures of down to -30°F. Once frozen, the contents and package are brick-hard and fragile. Ice cream is shipped in flexible material bundling multipackers. Loss of moisture during dry cold temperatures can desiccate the contents with adverse texture results. Temperature fluctuation during distribution and retailing can lead to leakage and to expansion of the contents.

Ice cream novelties cover a broad spectrum of portion-size products: sandwiches, chocolate covered bars on sticks, cups with frozen syrup, nut covered bars, and cylinders of water ice. Retail selling price per unit is low, so cost of packaging is an important consideration. Greaseproofness was formerly considered important, but this parameter disappeared when higher speed packaging machines appeared and were found to operate better with non-protective sheets.

Although attempts have been made to package cottage cheese in sealed flexible pouches for longer shelf life, this food remains in waxed paperboard or polystyrene tubs with snap lids. Sour cream generally remains in waxed paperboard tubs.

Yoghurt, a low-fat, high-acid product, which is a good yeast and mold growth medium, requires liquid tightness, acid resistance, and flavor impermeability.

Ripened cheeses and cheese foods are being distributed nationally by extending their shelf life through air exclusion. Absence of oxygen eliminates the possibility of mold growth and retards bacterial ripening action. This requirement has led to a specification for gas and moisture impermeability. The barrier must be greaseproof because of the fat content, and odorproof to prevent loss of desirable aromatics. Rather important is the ability to remove the packaging from the cheese, which can be a very sticky material. In order to retain the internal atmospheres, the package must be strong enough to resist puncturing. Opening and reclosability should be, but are not, requirements.

Cheese spreads are high fat and require protection against moisture loss. Opening and reclosure after use are not always required.

Dry milk solids have the same requirements as other dry foods, i.e., moisture exclusion. In addition, some consideration is given to oxygen exclusion by some packagers.

## Packaging

As described above, fluid milk and its analogues are packaged in glass, polyethylene extrusion coated SBS paperboard, and blow molded polyethylene. In Europe and Canada, and in test in the United States is flexible packaging of milk. Both major tetrahedral package machine manufacturers have attempted to introduce this type of packaging which is popular in Europe. Using polyethylene coated paper, these tetrahedral packages would require about a third less packaging material for a unit volume of contents than a conventional gable top paperboard carton. The space occupied by tetrahedral packages in retail display cabinets and home refrigerators has been an obstacle. The package has not been successful in the United States.

The polyethylene bag concept has been growing rapidly in Canada. Milk is filled into polyethylene tube stock of a special grade which can be reliably sealed at each end to create a pillow pouch. Roll stock is flexographically surface printed. Each pouch holds, in Canada, an Imperial quart. Three such pouches are multipacked in a tie-top reclosable polyethylene bag, the marketing package. The consumer places the pouch in an open-top pitcher, snips a corner of the pouch and is ready to pour. Surface tension of residual liquid closes the opened corner with finger pressure. After emptying, the polyethylene is removed, crumpled and discarded. Three mil polyethylene is a tough moisture-resistant material which, with its seal, can withstand squeezing, impact and low-temperature. It is an inexpensive packaging material in roll stock form. Packaging speeds can be increased when machines become available.[24]

Most ice cream is in rigid wax/ethylene vinyl acetate copolymer coated paperboard cartons or polystyrene tubs. Novelties may be in paperboard or plastic cups. Bar novelties, however, are packaged in flexible materials. Before automatic in-line form, fill and seal machines, these products were packaged in preformed glassine bags, the least expensive greaseproof bag material available. A few dairies used aluminum foil/glassine laminations and then aluminized glassine for bag stock to gain some graphic value. Today, an increasing proportion of bar novelties is packaged on automatic equipment in polyethylene extrusion coated MG roll stock in a pouch-style wrap. The packaging material and, of course, labor costs, are significantly decreased by this package.

Cream cheese, long packaged in a printed, embossed, aluminum foil lamination die-fold overwrap, has now entered the serving tub era. Although most cream cheese appears to be in the traditional package, formed tubs with foil lamination seals and polyethylene snap lids are now in the retail market. Tubs offer greater protection against moisture loss and reclosability, as well as a convenient package from which the contents may be dispensed.

Another group of soft cheeses formerly cut in wedges and saran wrapped are now in a thermoformed vinyl tray with a peelable aluminum foil lamination seal. The characteristic veins of blue cheese are reasonably well displayed in this package.

Soft spreading cheeses are often in reusable glass tumbler packages, the forerunner of today's injection molded plastic tub packages. They are also sold in aerosol cans. The chub package is now being used for soft spreading cheeses.

Processed cheese and cheese foods differ from the natural cheese from a packaging viewpoint, in that the latter is formed as a solid block that is cut or sliced for packing. Processed products are mixed, fluidized and poured into forms which may be the package itself or cast into sheets that become slices. Processed cheeses solidify on cooling.

In their formation, many natural cheeses are oiled, paraffined or otherwise coated. Some retailers cut cheese in the backroom for short-term storage. Generally, saran film or polymer-coated cellophane cling well to the cheese and can retard moisture and odor passage.

Longer shelf life is obtained by oxygen exclusion. When vacuum packaging was initiated, the slices adhered to each other so tenaciously that separating them became a difficult task. Carbon dioxide flushing proved to be a more practical mechanism for achieving a lack of oxygen without creating a negative pressure within the package.

Materials such as one-side wax coated cellophane and cellophane/polyethylene were and, to some degree, are still used for cheese packaging. The former is often used for processed cheese or cheese food. Neither packaging structure is an outstanding gas barrier even when properly sealed. Cellophane is prone to external abrasion breaks. Such structures as oriented polypropylene/polyethylene/PVDC/cellophane/polyethylene are expensive but offer gas, moisture and abrasion resistance.

An effective package form is a pouch. With large slices, the end seals are generally extended. End seals are folded over for compactness for small square slices which are stacked high.

To keep natural cheeses from adhering to each other in the package, paper interleaving is employed between slices. The same method is more difficult for processed cheese slices which are cast into sheets and which have greater tendency to stick together. Among the methods used, all of which are expensive, are individual slice pouch-style wrapping, folding polymer coated cellophane around the slice, and now casting into cast polypropylene.

None of the flexible packages for sliced cheese are as yet convenient to open and reclose. Individual slice packaging and complex outer structures add up to expensive packaging that still does not meet all the requirements.

Bulk natural and processed cheeses are also gas flush packaged in transparent films similar to those used for sliced cheese. Again, opening and reclosure are lacking. Some employ reverse printed cellophane/aluminum foil/polyethylene, a structure with good protective as well as decorative qualities.[41]

Dry milk solids are best packaged in three-or four-side seal pouches of paper/aluminum foil/polyethylene.

### Equipment

European machinery for polyethylene pouching of milk is vertical form, fill and seal equipment. Generally, an independently driven two-tube system, operating about 25 strokes a minute, is used. DuPont of Canada Ltd., one of the Canadian system marketers, uses a French Thimmonier unit. Quebec's Twinpack also uses a French unit. Neither has automatic pouch bagging. Hayssen has introduced an American vertical machine for the pouching function.

Automatic packaging for ice cream bar novelties is typified by a Hudson-Sharp Campbell horizontal pouch-style wrapper.

Anderson Bros. Mfg. Co. machinery is used for yoghurt packaging whether in preformed tubs or in roll stock rigid plastic. In-line thermoform, fill and seal packaging is performed on Anderson's imported Formseal equipment.

Much cheese packaging equipment is self-made by the packer. Smaller cheese makers extensively use the Hayssen RT horizontal form, fill and seal flexible packager. This machine has sliding dies which take into account the differences in size for positioning of the slices. Hudson-Sharp Campbell wrappers are also used for cheese packaging.

## DEHYDRATED FOODS

### Summary

Dehydrated foods include a growing body of both foods which are specifically dehydrated for preservation and foods which are formulated from naturally dry ingredients. Relatively few foods are dehydrated for preservation: instant coffee, mashed potatoes, soups. Most fall into the formulated category.

More recently, a considerable number of products have appeared in combination with complementary sauce mixes. Formulated mixes such as these require full moisture protection and fat resistance.

Three- and four-side envelope pouches formed from aluminum foil structures are usually used to package products such as sauces, beverages, and salad dressing mixes.

### Introduction

Dehydrated foods include fruits, such as prunes and apricots, freeze dried products which are used by astronauts during space travel, mixes formulated from dry components, and soups in which the constituents are dried separately and brought together in the package. Baking mixes are described in this section because they are dry products although most of the constituents are inherently dry rather than dried specifically for the function. Beverage mixes are manufactured by the same process, i.e., mixing ingredients which are dry in their natural state, and are also included here. Pastas, such as macaroni and spaghetti, and their mixes; rice and rice mixes; beans; and ice tea mixes are also in this classification.

Common to most of the products in the grouping is addition of liquid by the consumer to prepare the final product. However, dried fruits are often consumed in their dried state.

The terms dehydrated and dried are synonymous, although some would define dehydration as the controlled process of removing water, and drying as a process of allowing nature to perform the operation. In either case, the objective is to reduce the water content to a level below which microorganisms do not grow and enzymes cease to function.

Two basic methods of drying exist: adding heat to evaporate the water, and adding heat to sublime ice as in freeze drying. Heating wet food changes its characteristics and may drive off desirable components with the water. Reducing the pressure above the food lowers evaporation temperature of water and decreases heat damage due to the process. Freeze drying imparts the least change to the product as a result of the process.

Few products, even under the most careful drying conditions, escape some processing change. For this reason many new dry foods are being manufactured by using as many naturally dry materials as possible. These materials include sugar, corn syrup solids, flour, citric acid, and starch.

Since raisins, prunes, and apricots are sweet acid products, partial drying concentrates the sugar and decreases the pH to a point where little microbiological action can occur.

Rice is a milled and cleaned grain. Some is parboiled to render it more conveniently rehydratable. It may also be completely cooked and then dehydrated to create so-called quick (instant) rice. Pouches of sauce may be included to flavor the rice, or seasonings may be actually mixed with the quick rice so that all will rehydrate together upon addition of water. Basic rice is a commodity; converted rice, and quick rice and rice mixes are products; all are distributed and marketed as groceries.

Pastas, including noodles, are made by extruding a special flour water dough into the desired shape and drying slowly to obtain moisture loss throughout the structure. Most of these are marketed as commodity items, with few differentiating characteristics among brands. Many local manufacturers exist who produce shapes for specific market segments. There are few strong regional manufacturers.

The basic constituent of cake mix is flour. Dry sugar, flavorings and leavening are mixed with the flour and shortening. One manufacturer claims to wet the mix and subsequently spray dry it. Today cake mixes are close to being commodities despite intensive efforts by the national brands to market them. Frozen and locally baked sweet goods have eroded market shares.

Beverage and cocktail mixes are mixtures of sugar, citric acid and flavorings, which can be and are mixed by almost any contract packager with a pouch machine. They are simple to make, but not too simple to market as there are definitely strong brand loyalties. There appears to be considerable seasonality to beverage mixes, with summer being the peak period.

Salad dressing mixes, another popular item, are basically salt, seasonings, and spices brought together in one place to season oil, mayonnaise and/or vinegar which the consumer supplies. Manufacturers may even supply a jar in which to mix the dressings or a disposable bowl in which the ingredients are mixed.

Sauce mixes and omelet seasonings are basically the same, an appropriate mixture of dry ingredients to which the consumer adds basic ingredients. The mix is actually a flavoring element. All of these can be blended by any competent packager; manufacture is simple; skill rests with the marketing.

In dehydrated soup mixes, soup vegetables and other components are dried by hot air and then combined in the final dry product. Noodles may be added to some of the soup mixes. Potatoes are cooked, mashed and dried on heated drums and comminuted before packaging to produce dehydrated potatoes.

Some freeze dried shrimp, meat and fish have been introduced to the HRI market. Mayonnaise may be added to freeze dried tuna fish to obtain reconstituted tuna fish salad.

Iced tea mixes may be dried versions of tea beverages, often with added flavoring and sweetening.

This category will expand as more products are formulated from the many dry ingredients now being made available to the food industry. A decided trend is under way toward creation of new products by mixing individual elements as opposed to processing conventional foods to preserve them.

### Packaging Requirements

Obviously, the principal requirement for dehydrated food packaging is the exclusion of water vapor. Water rehydrates the product and makes it susceptible to the same spoilage as the whole food. Water in small quantities creates conditions which accelerate biochemical and even enzymatic action, and can form substrates for mold growth.

Dehydrated foods, if processed at low temperatures, may contain active enzyme systems which are operative as moisture and temperature rise. Dry foods are concentrated biochemical systems in which activity continues. Among the more active systems is the oxidative process which can lead to rancidity, browning and other off-flavor and color phenomena. These reactions are accelerated as moisture content increases.

Exclusion of oxygen is a desired but infrequent practice. Gas flush or vacuum packaging is sometimes employed, but usually atmospheric packaging is used despite the definite flavor, color and texture deterioration that occurs.

Dehydrated solids are hard, sharp-cornered and abrasive materials which can easily cut, scratch, and otherwise damage packaging material. Abrasion and breaking can lead to powdering and release oils that can react with packaging materials.

Because they are concentrated by elimination of water, dehydrated foods have a high value per unit of weight. Fill control can be difficult especially with powders, many of which have low densities.

Grain products, such as rice and pastas, and beans contain native starch, and so are difficult to rehydrate. Thus, they are not especially susceptible to moisture or other environmental factors. Once parcooked, however, they become, like other hydrated starch products, far less stable.

### Packaging

Fruits that have been sun dried are handled like produce, i.e., in paperboard cartons. With waxed paper overwraps, moisture protection is sufficient for marketing purposes. Some fruits, such as raisins, are packaged in portion packages and multipacked in cellophane. Others employ aluminum foil lamination overwrap to obtain the dual benefit of decoration and moisture protection. Some may be purchased in transparent cellophane or cellophane/polyethylene pouches. Many fruits are dried to a specific moisture level, low enough to prevent spoilage but high enough to insure softness for ready usage. Visible packaging is today better suited to the produce than the grocery department. Cellophane can be an adequate packaging material since its moisture content will remain high as a result of the package contents. The dry fruit softness helps to allow the film to conform without breaking or abrading.

Rice, pasta and bean packages are unlined, printed paperboard cartons; windowed paperboard cartons (typical for pasta to show the product); or paper or film bags. Film bags for rice or beans are often preformed cellophane despite the problems which arise from physical damage of this relatively breakable film by the contents.

One major pasta manufacturer has achieved success by packaging noodles in well-designed duplex cellophane bags. Basic lines employ a window in the printing to display the product. Several products in the line are packaged in 100% reverse printed photographic reproductions of the product ready for eating.

Quick rices are fairly insensitive, and so are packaged without liners. When a sauce mix is included, however, the mix is treated like a dry food in a four-side seal lamination pouch. To keep printing inks from exposure to the rice (or pasta in the case of a pasta dinner) an outer transparent glassine ply is printed prior to laminating to the aluminum foil. Although cellophane could be used, it is more expensive and not needed for attractiveness on the interior of an opaque paperboard carton.

Cake mixes are not totally moisture insensitive; however, grease is a problem. The shortening should not stain the highly decorated carton. Most of the major cake mix manufacturers use glassine liners to provide greaseproofness and, to some degree, moistureproofness. One manufacturer uses bag-in-box constructions with transparent polyethylene bags. Angel food cake contains dried egg albumin, a material which must be treated like any other dry food. Glassine/aluminum foil laminations are often employed, again because of the presence of flour contents.

Beverage mixes, such as fruit-flavored drinks, are packaged in single-wall constructions such as paper/PVDC or the paper/foil combinations for dried foods.

Other dry powders, beverages and sauce mixes use paper/polyethylene/aluminum foil/ polyethylene in four-side seal envelope pouches. There are numerous variations on the basic construction, which employs aluminum foil as the moisture, gas, oxygen and fat barrier. Paper may be adhered by adhesive rather than by polyethylene. Because some dry products, such as noodles or vegetables, are sharp-cornered and abrasive, there is some use of tougher ionomer in place of the interior polyethylene.

Attempts to use paper/PVDC combinations in the past for products like iced tea mixes met with failure because PVDC, despite its virtues, is not a moisture barrier like aluminum foil.

Relatively simple structures are used for dehydrated potatoes. When these products were introduced, the sophisticated packaging materials of today were not available. As a result, they entered unlined paperboard cartons, a package from which they have not really graduated.[15]

### Equipment

Dried fruit is packaged on vertical form, fill and seal equipment. Preformed bags are doubtless used by some small operators, but automatic equipment is today a labor-saving mechanism, inexpensive enough for most businesses. The significant component of any equipment used for dry foods is the filler, usually purchased separately since the basic flexible film handling mechanism can be used for a range of products. Dried products are generally low density and valuable materials. Fillers, there-fore, must be able to move precise quantities into the pouch without damaging the food or interfering with the seal. Dried fruits are somewhat sticky and can get hung up in a gravity feed.

Rice and beans still appear in manually filled preformed pouches, heat sealed or even sealed with transparent tape. This type of package has no headspace and so is filled to complete capacity. Pasta products may be flowable noodles or macaroni in which case they are filled on vertical equipment using gravimetric fill. Triangle and Woodman slant machines are often used for this purpose. Long fragile pastas such as long strand spaghetti, if not cartoned, may be manually bundled in cellophane on Hobart Corley-Miller wrappers which complete the wrapping by adhesive sealing the ends.

Cake mixes in lined paperboard cartons are packaged on Pneumatic Scale Corp.'s double package makers. Those in bag-in-box configurations use conventional vertical form, fill and seal equipment for regular mixes. Egg white solids in aluminum foil lamination pouches may be packaged on horizontal form, fill and seal equipment such as Bartelt or Delamere & Williams.

Generally, these two pieces of equipment in one or another model are used for most three- and four-side envelope pouch packaging in the United States. A few Höller machines are employed for beverage mixes. This equipment handles stiff sheets well without damaging them. A single sheet is folded in half and a three- or four side face-to-face weld seal is made. The envelope is opened, filled and sealed shut. Machine motion may be intermittent or continuous so that speeds of up to 300 pouches per minute can be attained. These pouches have limited capacity because of the shape, and seal areas can become wrinkled. Further the amount of expensive packaging material needed per weight of contents is relatively high. Nevertheless, dry foods are usually sold in small portions, and the pouches are most satisfactory. Furthermore, foil's dead fold features allow for reasonable reclosability.

Horizontal form, fill and seal equipment may be rigged for gas flushing. Vacuum packaging may be more difficult, but the structures used for dry foods generally do not withstand vacuum.

# DESSERTS AND DESSERT MIXES

## Summary

Dessert mixes include gelatin, starch, pregelatinized starch, and rennet products. Because there is a long tradition behind the products and their packages, they are generally flexible materials inside paperboard cartons.

Glassine, waxed glassine and waxed glassine laminations extrusion coated with polyethylene are typical structures for the pouches in the cartons.

Flavored freezer liquids are in liquid-tight, low-temperature resistant four-side heat seal pouches.

## Introduction

Despite the apparent calorie consciousness of the American citizenry, sales of sweets continue on the upswing. Sweet foods are not only consumed at meals but as part of the now common fourth and fifth snack meal. The dessert category in this discussion includes gelatin and starch pudding mixes, and some flavored liquids intended for package freezing and consumption. Gelatin and starch pudding dessert mixes are classics of the genus and are among the earliest of the package groceries.

Gelatin dessert is a protein with added sweetener, acid and flavoring in powdered form. Encapsulation of flavors a few years ago led to elimination of the need for greaseproofness. Formerly, flavoring materials were oily and dispersed on the powder. A broader range of natural and synthetic flavors is now being used, and further, artificial sweeteners are being used for the dietetic versions.

Native starch puddings have been expanded into precooked starch puddings, requiring only the addition of milk and mixing for reconstitu-

---

## TABLE 2

## IMPORTANT MATERIAL COMPONENTS FOR THE IMMEDIATE FUTURE

| | |
|---|---|
| Nylon: | Boil-in-the-bag and vacuum packaging laminates for thermoform-fill operations. |
| Aclar: | Transparent packaging for materials requiring the greatest protection from moisture vapors-pharmaceuticals, freeze dried products. |
| Ethylene-vinylacetate copolymers: | Stress crack resistant laminating webs, heat seal medium for laminates with low temperature sealability: bag-in-box laminates. |
| Ionomers: | Oil and grease resistant layers, low temperature heat sealing layers, heat sealing to glass and to metal. |
| Polypropylene: | Oriented:   Outer layers for vacuum pouches form-fill packaging.<br>Cast:   Inside layers for sterilizable packs. |
| Vacuum deposited metal barriers | Form-fill packaging, deep frozen products, light resistance, attraction, machinability. |
| Adhesives for sterilization: | Sterilizable flexible packs. |
| Polyvinylidenechloride: | Best barrier coating yet found for moisture and oxygen. |
| Source: | Mr. E. V. Southam, Director, Dickinson Robinson Group Ltd., Bristol, England, *Packaging Technology*. |

tion. With other additives, aerated versions may be prepared. Some are especially formulated to be pie fillings. Rennet puddings are sold in both powder and tablet form. The dry mixes are distributed as dry groceries, i.e., wholesalers and warehouses under ambient conditions.

Frozen suckers may be purchased in liquid form for home freezing or in the frozen form. Both are packaged in the same multiple flexible package.

### Packaging Requirements

Gelatin desserts are susceptible to moisture pickup and caking. Moisture entry should, therefore, be restricted as much as possible. As indicated above, greaseproofness is not of any great consequence at present. Retention of flavor aromatics is, however, of great importance.

Starch puddings, particularly those containing pregelatinized starches, can deteriorate by moisture absorption. Again, flavor retention is of consequence. Rennet puddings in tablet form are highly susceptible to moisture degradation and so should have complete moisture protection.

Flavored liquids must, of course, be in liquid tight packages. Flavor retention is important, and for marketing reasons the packaging is transparent.

### Packaging

Gelatin and starch desserts have a long history and, as a result, have a packaging tradition. The products were packaged in paperboard cartons with die-folded, waxed glassine liners. The glassine served as a base for wax to provide moistureproofness, and acted as a fat and flavor barrier. Effective high-speed heat sealing was not available and the die fold was developed.

With the decreased need for greaseproofness and the increased need for moistureproofness, the interior lamination has undergone changes. The paperboard carton has remained. At present the most widely used carton liner is one-side polyethylene coated wax-laminated glassine. This may be in a preformed pouch or, more commonly, a pouch made on form, fill and seal equipment and automatically inserted into paperboard cartons. A limited amount is still packaged in die-fold glassine or paper.

Rennet pudding powder is in die-folded envelope interiors. The chipboard carton is overwrapped in a printed foil/wax/strike-through tissue lamination. Rennet tablets are now packaged in pharmaceutical unit-packages, using heavy-gage aluminum foil laminations.

Freezer liquids are in a number of laminations with polyethylene interior extrusions. To insure against potential breakage due to internal liquid expansion or fracturing due to freezing temperatures, a tough low temperature outer-ply such as polyester is employed. The package is a four-side sealed tube with wide face-to-face seals to assure liquid retention.

Gelatin desserts are among the few packaged products which are shipped in flexible containers rather than corrugated. By tight bundling in kraft paper, a compact cellular structure capable of stacking is obtained.

### Packaging Equipment

As in so many other packaging problems involving dry foods, the package is a pouch formed on horizontal form, fill and seal equipment. Those packages having internal polyethylene extrusion coatings are typically manufactured on Bartelt-type equipment. These machines included automatic carton sleeve set-up and filling.

Liquid freezer items described in this section are typically packaged on Circle equipment filled with a positive displacement pump. The machine works from two wide rolls which are slit into narrow packages during formation.

# FROZEN FOODS

### Summary

The frozen food industry, although large and growing, has not even begun to approach expectations. Because of ease of entry into the business, the low cost of raw materials and the fact that most packers hire the same outside distributors, a highly price competitive environment has developed. This situation has led to two extremes—a large number of commodity fruits and vegetables sold on a price basis regardless of quality, and, at the other end, a large number of highly specialized foods, again often with marginal quality. Neither extreme has helped industry expansion. Into the void have entered a number of higher quality product lines such as baked goods; but, again, low cost imitations have injured the quality image.

Freezing enables a packer or marketer to prepare and sell almost every type of food. It is especially suited to precooked and multicomponent products such as dinners.

Frozen foods should be packaged to retard moisture and volatile aromatic losses, to exclude oxygen and light, and to prevent the hard foods from abrading and damaging the packaging. Because low temperatures retard deteriorative changes, there is a tendency to pay little attention to protection requirements.

For several reasons, possibly centered about the shape irregularities and the need for superb graphics, frozen foods are generally packaged in paperboard cartons. Some continue in waxed paper overwrapped cartons, but this, like so many other early frozen food packages, is declining.

The advent of individually quick frozen foods brought with it the large polyethylene bag containing loose pack frozen food. A substantial, but not major, portion of IQF product is in this economy-oriented package.

The convenience route to frozen food marketing has been highlighted by prepared meals and meal entrees. Boil-in-bag packaging represents a convenience package for higher quality meal components. Vacuum packed pouches are used for this product, often packed as a two component system with solid and sauce separate.

There is a dearth of good frozen food packaging equipment and, thus, innovative frozen food packages.

## Introduction

As a result of post World War II consumer purchasing of refrigerator-freezers, the American frozen food industry enjoyed a rapid expansion. Ice cream, frozen vegetables and fish and then frozen concentrated citrus juices grew rapidly in separate waves.

When precooked frozen foods made their debut with French fried potatoes and fish sticks, there were great expectations and predictions for frozen foods. All types of food products were frozen and placed on the market. Frequently, product quality was a secondary consideration. Distribution handling practices

were (and still are) poor, leading to temperature fluctuations and even thawing, thus degrading quality. Because almost anyone with a stove and a home freezer could become a frozen food entrepreneur, a multitude of marginal enterprises entered the business. As a result of low quality arising from price competition and from the other factors, the consumption of frozen foods did not increase as much as expected.

Even today, after a considerable shakeout of marginal operators, residual effects remain. Further, the distribution system, despite radical improvements arising from legislative threats to impose 0°F maxima, is still sadly short of even minimum standards. Private labelling is a rule rather than an exception in the industry. Introduction of high quality items has been followed hard by lower-priced competition. There is little sophistication in formulation, processing or freezing of products.

In spite of the obvious deficiencies, frozen foods have gained a position among American food purchases. The industry is growing, and even per capita consumption is increasing because of the compelling advantages presented by the products.

Older products have broadened their bases. Frozen concentrated juices now include lemonade and drink concentrates. Toaster items evolved from frozen prebaked products. The precooked frozen dinner is part of the American legend, if not the dietary.

Frozen foods may be categorized into fruits, vegetables, vegetable plus sauce, precooked meals, juice concentrates and specialty foods. Frozen fruits are very much part of the commodity area of the frozen food business. Generally, they are sugared, with the raw material being high-acid and highly colored. Berries are marketed in two ways—as small packages with syrup and in bulk so that the desired amount may be dispensed for serving.

Vegetables are another commodity of frozen food business. Green vegetables are partially cooked by blanching prior to packaging and freezing. The largest single group is French fried potatoes which are fried prior to packaging and freezing. Both classes are packaged in small unit packs and, in bulk, in reclosable bags.

Vegetables have partially moved out of the commodity area with the successful reintroduction of the boil-in-bag pouch. Attempts to use the package to hold blanched vegetables for consumer convenience met with resistance perhaps because frozen vegetables were fairly easy to prepare. When sauces, such as butter or cream, were combined with the vegetables, however, the commodity was upgraded and a consumer convenience was introduced. The flexible boil-in-bag pouch proved cheaper than previously proposed rigid aluminum foil packages.

Main courses are generally marketed as oven preparation products in aluminum foil containers. Even more recently, straight-sided aluminum foil tubs have even been used so that the package which is also a baking dish may be used as a serving dish. By using decorated aluminum foil, main course packages have taken on the function of the rigid aluminum tray of the frozen meal.

Expansion of frozen foods in terms of new product introduction has been in precooked meals, meal components and specialties. Pot pies which act either as meal components or as full meals, macaroni and cheese, fruit pies and pizzas now in numerous forms, and lasagna are but examples of food frozen in metal containers for oven reheating. The ethnic food fashion must be a reflection of activity in this sector of the frozen food business. No other preservation process is capable of capturing and retaining as much of the original flavor, color and character of the original product as freezing. Thus, a flood of new products results from the pairing of freezing with nationality products.

The range of specialties in frozen foods is broad. In addition to meals of various national origins, freezing is used for portion-size precooked fish, meals, and snacks; baked goods; unbaked doughs, toaster items (e.g., waffles) unstable at room temperature; portion control meats; partially cooked meat; and fabricated foods such as French fried potato analogues. Almost every food cooked prior to eating is being frozen commercially.

Frozen concentrated orange juice is a commodity which altered the consumption patterns of the country. Once the concept was accepted as a matter-of-fact item on the market, it multiplied. New sizes became commonplace, and then the product expanded into other juices and fruit-flavored drinks.

Frozen foods require specialized storage, distribution and retailing channels. Further, most frozen food manufacturers are so small that they do not have their own distribution and marketing. Product is out of their control during most of the cycle.

## Packaging Requirements

Fundamental to all packaging for frozen foods is the ability to withstand 0°F and lower temperatures. Many materials can embrittle and shatter at the least shock at these temperatures.

Good technical and business practice indicates that 0°F is the maximum temperature which frozen foods should attain during their lives. In actuality, many must undergo considerably lower temperatures as a result of direct contact with refrigerant plates, or cold air blasts, or product previously frozen cryogenically. Foods may be packaged before or after freezing.

This means that some foods, such as blanched vegetables or precooked foods, may actually be warm or hot during packaging. The more common practice is to chill the food after heating and before the next step, but this cooling may not be complete.

Moistureproofness is an important, but not critical, requirement of frozen food packaging. Frozen foods can lose considerable moisture by sublimination due to cold, dry environments which leads to freezer burn. Freezer burn damage is very visible, and packagers are conscious of the problem. Somewhat more critical, however, is the formation of cavity ice inside of the package due to temperature fluctuations.

Many precooked foods are deep-fat fried prior to freezing. Packaging for such foods must be greaseproof. Other foods such as hamburger patties and vegetables in butter sauce contain fat not derived from frying.

Oxidation is a source of deterioration in many foods. Fatty fish such as mackerel if not protected against oxygen can turn rancid. Color loss of vegetables such as carrots and fruits such as strawberries may be partially due to oxidation. Because of the low temperatures involved, however, oxidation is a slow process. Exclusion of oxygen is not often an objective of the packager unless the problem is extremely critical. This is the case only with such foods as poultry and boil-in-bag products. Boil-in-bag packages are evacuated to assure against internal swelling and bursting during the cooking part of reconstitution. The vacuum also helps to keep the product submerged beneath hot water during heating.

Packages that are intended for heating must, of course, resist that heat. If submerged, the seals must remain intact to insure against loss of contents and against ingress of heating water. The moisture of oven-heated foods should be retained by the packaging during heating.

Many of the desirable bright colors of fruits, vegetables and meats are subject to alteration as a result of exposure to light. Fat rancidity can be accelerated by light. On the other hand, some marketers believe that frozen food visibility is a desired merchandising tool. Due to internal frost, visibility within packages is difficult to achieve, and so transparent frozen food packaging is debatable.

The presence of liquid with frozen foods dictates that the packaging be liquid tight. Sugared fruits, however, also carry considerable syrup. The presence of sugar means abrasive materials in contact with the package interior.

Frozen foods are often hard and have sharp corners, and so may inherently have the ability to damage packaging of insufficient strength. Frozen foods, although hard, may not have good compressive strength. Packaging must, therefore, protect the product against dynamic and static stresses.

Some of these stresses may be rather extreme as in the plate freezing of some vegetables and fish fillets. Plates containing refrigerants are brought into high pressure contact with the package. Individually quick frozen (IQF) foods are frozen outside of the package and can act like small missiles when being filled.

Fruit may be packaged as IQF or in containers as blocks with sugar and/or syrup. Vegetables may be packaged IQF or, after chilling, into containers for plate or air blast freezing. Boil-in-bag packages, because of the potential for microbiological spoilage if held above freezing, are frozen individually if possible. Platters and entrees should be frozen before casing, but this is not always the practice. Rather, pallet freezing is not uncommon. Product after packaging and casing is palletized

before freezing. Large pallet loads are then subjected to still air or air blast freezing. The time period for freezing, especially for the center, can run into days, so the packaging must indeed be liquid tight.

Although some packaging which dates back many years is still in common use for frozen foods, there is a definite consumer consciousness. Convenience of opening, dispensing and subsequent use has been an issue for packaging frozen foods in recent years. More than any other food product category, frozen food packaging has been the objective of obtaining better decoration. This may be due to several factors, among which are the inherently poor appearance of frozen foods in the frozen state.

Although the distribution system with all of its problems has not changed greatly, there are several innovations under way at the retail level. Open-top reach-in display cases are giving way to shelving with packaging at eye level. These new modes of display will dictate that the form and graphics positioning on frozen food packaging be altered.

### Packaging

The first successful commercial frozen foods were in waxed paperboard cartons overwrapped with printed wax paper. The cartons provided some structural protection and, with wax coatings, prevented the product from sticking to the paperboard. Waxed paper in a double point end fold overwrap provided moisture protection and a medium for graphics. This package is still used for many frozen vegetables.

Although used by many packagers, the overwrapped carton has generally been supplanted. Printed wax/ethylene vinyl acetate copolymer blend coated hot-melt closed paperboard cartons are now common for many IQF and wetpack frozen foods.

A substantial part of IQF products is bulk packaged in one- to five-pound polyethylene bags. Generally, these bags hold one or two pounds of product. The question of visibility is underscored here. Some packagers, conceding the haziness of polyethylene and the poor appearance of frozen foods especially after some temperature cycling, employ printed opaque polyethylene. Others, contending that visibility aids in connoting economy, employ transparent bags with minimum decoration. Polyethylene is a good moisture barrier, is easily heat sealable, has good low temperature durability, and is inexpensive. Bags are made of approximately 3 mil polyethylene. A number of them now are manufactured with ties for reclosure. The intent is for the consumer to use part of the contents and then return the package to the freezer for use at a later time. The polyethylene bag is generally a simple weld seal tube type. A few packers use a paper/polyethylene laminate, but this is subject to wetness problems.

Boil-in-bag constructions, having far more severe requirements, are considerably more sophisticated. Flexible packages for boil-in-bag foods are four-side seal pouch-style envelopes generally fabricated from polyester with low or medium density polyethylene interior extrusion. Polyester imparts strength and barrier and is heat resistant. Polyethylene adds to the moisture barrier and provides heat sealant for the pouch. Medium density polyethylene is more suitable because of its higher melting range and its greater resistance to oil penetration. Some constructions use PVDC to assist in vacuum retention.[40]

Aluminum foil laminations have been proposed and used, but because cost of protecting the foil is so high, this is not a widespread practice. Some of the nylons have been proposed; polypropylene with a polyethylene extrusion has also been proposed.[15 40]

Laminated film pouches are now used for thaw-in-bag fruits in syrup. Constructions are of the same nature as boil-in-bag, with both freezing and warm water temperatures being consequential.

Some of the frozen toaster products are packaged in tie-top reclosable polyethylene bags with paperboard containers. Frozen waffles would be an example of this application.

Frozen dinner trays are almost all topped with printed unmounted aluminum foil crimped under the edge. This foil is easy to remove after heating, and helps to hasten heating by creating internal steaming during heating. Foil also retards moisture loss in storage. A variation on this system employs polyester film as the top seal. This may be peeled back after heating.

Several years ago, there was an effort to introduce frozen concentrated juice in tetrahedral packages. Formed from paper/polyethylene/alu- minum foil/polyethylene these packages suffered from the same drawbacks as other tetrahedral packages and so were withdrawn from the market.

A few specialty fish items are marketed in flexible packages. For example. shrimp may be in polyethylene bags. Whole frozen fish may be in flexible film wrapping.

Inexpensive pizzas and similar locally prepared specialties may be wrapped in polyethylene, an inexpensive means for a small operator to package product. In Canada, precooked fish is placed on a tray, and a shrink film is placed on the top and shrunk on to the package. This is a merchandising tool—making the factory-prepared fish appear to have been cooked nearby and then quickly frozen.[15 18]

**Equipment**

The frozen food industry has not been generally noted for packaging innovation. Starting with fruit and vegetables, they were able to take advantage of canned fruit and vegetable technology and processing equipment. Thus, package filling machinery stems from the canners. Carton handling equipment is conventional and, oddly enough, fairly uniform from plant to plant.

Unprinted paperboard cartons are mostly lock-corner types, automatically formed, or infold types for maunal set-up. Filling is automatic only if the product is flowable as IQF peas. Otherwise, and this is more the rule, the product is filled manually or semi-automatically. Fish fillets or fish sticks, for example, are laid in by hand. They are not sufficiently uniform to be counted to a given weight as they have a highly irregular shape and form.

Overwrapping equipment is standard waxed paper machinery such as Package Machinery Corp. or Hayssen units, both of which are in fairly common use in plants employing the overwrap system.

Bagged IQF fruit, vegetables and meats are packaged on vertical form, fill and seal equipment. The most widely used machinery is the Hayssen twin-tube vertical unit. Because the product flows, volumetric fill is employed. A three-side pouch-style bag is formed, filled, and sealed and dropped out of the unit. A problem of enclosed air can lead to pillowing, so the equipment has a device to squeeze out some of the excess air. Pillowed bags, even made from polyethylene, can burst from pressure.

Boil-in-bag foods usually contain two components, the solid and the fluid. Each is frequently added separately to the package to minimize adverse effects in processing and to assure that the appropriate amounts of each are in the consumer pouch. A considerable amount of boil-in-bag product is evidently still packaged manually using preformed pouches. Two types of equipment are employed to effect this packaging; both are horizontal types, but one has horizontal feed and the other vertical. The latter is typified by Bartelt equipment, often with two-station filling for two component products. There is still the problem of limited amount of volume that can be filled without distorting the pouch and the seal area. Vacuum is not easy to draw in four-side seal pouches of this nature as opposed to the horizontal feed machines.

Processed meat packaging equipment has found some use in packaging boil-in-bag foods. Two films are used, a top and a bottom. The bottom film is heated and vacuum drawn prior to one- or two-station filling. A top layer of film may be sealed in a vacuum chamber, or by means of a needle withdrawing air during the sealing operation. Standard Packaging Corp. 6-9 is capable of 50-60 packages per minute, and its 6-14 can attain speeds of up to 90 packages per minute. The other widely used equipment is manufactured by Mahaffay and Harder or by Royal Vac.

Foil trays are packaged manually with mechanical assist for flowable items like mashed potatoes, gravies and spicing. The foil is crimped on automatically from roll stock on equipment whose design originated with the aluminum foil companies.

Precooked food packages with shrink tops are an innovation with equipment made and supplied by Cryovac. Filling is manual into denested trays. Film is from roll stock which is formed and sealed onto the product and tray by a flexible metal mesh that conforms to the shape. A shrink tunnel completes the job.

Unfortunately, little has been developed uniquely for this industry. With such a small portion of frozen foods in flexible materials, not much attention has been paid to the expansion of this method of packaging. As a result,

IQF packaging uses stock vertical equipment found in many other parts of the food industry, and boil-in-bag equipment was borrowed from the processed meat industry.

This problem will continue in the future as the frozen food industry continues on its highly segmented route. Because of the relatively small and seasonal volumes (shaved somewhat by IQF systems), the lack of adequate packaging equipment will become more acute. This leads to a uniformity of packaging from company to company that reduces package differentiation to practically nil. Total reliance is on graphics which compete loudly. That so young an industry could have fallen into this trap before becoming firmly established with its own packaging is inexcusable.

# MEAT

**Summary**

Retail meat packaging is divided into two distinctly different categories: fresh meat and processed meat. The former has not been rationalized. Processed meat has become a branded, packaged product.

Fresh meat continues as a commodity, brought to retail stores in carcass or primal form, and cut and packaged in supermarket backrooms. The methods must necessarily be variable in sanitation as well as efficiency. Retailers believe that red meat color and quality can be retained only with cutting close to final consumer sale.

Fresh meat packaging is for appearance. Tray and film overwrap systems are almost always employed. Films are selected for their clarity, ability to pass oxygen required for color maintenance, and lack of fogging. Moisture retention for weight control is another important requirement. PVC in stretch and shrink forms is most often used. Much equipment is manual or semi-automatic because the volume does not justify high speed equipment.

Centralization of fresh meat packaging must come because of labor costs and inefficiencies of small back-room operations. Good preservation can be achieved by improvement of present practices. Atmospheric control typified by chub packaging can assist. This latter demonstrates that color may have little bearing on consumer acceptance.

Processed meat has taken advantage of many of the innovations in flexible packaging and, as a result, there has been nearly a complete conversion to branded, convenient, and functional retail products. Processed meat color is stabilized by curing agents which also help retard microbiological growth. As a result, oxygen-free packaging can be used to effect long distribution channels from central factories. Vacuum and gas-flush packaging are common. Vacuum packages generally consist of drawn flexible material conforming to the meat shape and size. The thermoformed sheet may be either polyester/polyethylene or nylon/polyethylene. Since these sheets are somewhat expensive, flat cover sheets are less expensive materials such as PVC/PVDC/polyethylene.

Recently, rigid PVC and XT polymer have been employed for deepdrawn packages for both luncheon meats and frankfurters.

Carbon dioxide gas-flush packaging has been used by medium-size packers for loose packed sandwich meats. Laminations containing PVDC for gas barrier are used on these three-side seal, horizontal form, fill and seal pouches.

## Introduction

In the midst of dynamic retail merchandising stands fresh meat almost unscathed by progress and pressure. Centralized packaging, the name applied to a system long accepted for milk, pie, and ice cream, is apparently regarded as an ultimate goal. Whether viewed as a specter or a hope, removal of fresh meat packaging from the retail store is inevitable.

In contrast, processed or cured meats, typified by frankfurters and luncheon meats, have emerged from commodity class to become a family of products. Packaging has been a vital contribution to this marketing emergence. Packaging of the two are sufficiently different that they may be treated separately.

## Fresh Meat

At present, animals are killed, dressed and broken into carcasses at meat packing plants near the livestock feeding area. Carcasses chilled to below 50°F may be shipped, hanging, to the retailer or to his warehouse for subsequent reshipment. Sides or quarters, unpack-

aged, are still the most frequent raw material shipped into the back rooms of retail stores for breaking and cutting into retail cuts.

A less frequently employed method, but one which is growing, is breaking the carcass into primal or subprimal, i.e., knife-and-saw-ready, cuts and shipping this far less waste-containing component to the retail back room.

Almost all reduction to retail cuts and consumer packaging occurs in the back rooms of retail stores. There are few federal, state or even local regulations or standards governing the operation of these more than 30,000 establishments through which almost all fresh meat is processed. Each chain or individual manager operates each back room according to individual experience or needs. There are really no universally accepted guidelines for this management, nor any widely accepted totally or partially integrated equipment systems.

It stands to reason in a situation like this that, at the very least, there would be wide variations in operations, sanitation standards, refrigeration, labor costs, and packaging methods. What little data has been gathered for published studies indicates that there are indeed broad ranges of cleanliness, temperatures in holding and display coolers, and effective employment of personnel.

Back-room butchers, who double as retail clerks, break and cut the entering or stored sides, quarters or primal cuts into retail cuts at ambient or slightly lower temperatures. These consumer-sized pieces are then placed in pulp, foamed polystyrene and transparent oriented polystyrene trays. All consumer packages today are hand-packed in a small number of standard size trays, most of which do not conform to the meat cut size. This permits ease of subsequent wrapping but, perhaps more important, allows the plastic molders to compete vigorously on a price basis.

Filled trays are subsequently overwrapped with a transparent film, the combination of film and tray being allegedly cheaper than a single folding carton. The claim is made that the consumer must see the meat or else she will not make a purchase.

Clear films are generally used and, as a result, red color is an all-encompassing quality objective to be attained, even at the expense of other characteristics.

Since back rooms often do not generate sufficient volume to justify mechanical or automatic equipment, machinability is not always a consequential factor. Most back rooms today probably make use of mechanical assist for manual wrapping, and some semi-automatic equipment which cuts off the sheet to standard sizes allows the operator to wrap and then effect a tack seal. A good heat seal would be difficult to make and difficult for the consumer to open. The wrapped tray may then be automatically weighed and labelled although only rarely with a brand or store identification analogous to cookies or bread.

All of this packaging is performed at ambient or slightly below ambient temperatures under varying sanitation conditions. Workers have varying degrees of training and generally a lack of appreciation of the microbiological implications of their activities. Most back-room operations are reasonably clean, but few would meet the scrupulous sanitation standards of a dairy or bakery. The back-room operation, small volume in terms of meat, may be large by other food plant size references. On this basis alone, back-room operators should adhere to standards of food plant operation.

### Fresh Meat—Packaging Requirements

Fresh meat, used by consumers to prepare steaks, stews, roasts and ground beef dishes, is derived from the tissues of slaughtered animals. Only the life processes of the animal are destroyed in the kill; the tissues remain biologically viable and subject to all reactions inherent in such materials.

Beef is principally protein, with some fat, a small amount of carbohydrate and varying quantities of vitamins, minerals and other trace materials. In life, the protein has been formed from amino acids ingested or manufactured by the animal. With no life controls or amino acid input, the direction of biological reaction is toward reduction of protein into components.

Since beef is one of man's most flavorful and prized means of supplying nutrition, enzymatic degradation of meat should be limited. The reactions involved are temperature dependent, more than doubling in rate with each 10°F increase in temperature up to about 130-140°F, when denaturation inactivates the enzymes.

Proteins may also be altered, although to a far lesser degree, by biochemical reactions which are not enzymatically catalyzed. These reactions occur because of proximity of reactive molecules to each other. Reaction rate is also temperature dependent, approximately doubling with each 10°F increase in temperature until there are no more reactants available.

Additionally, the fat components of beef which contribute desired flavor, tenderness and calorie value are subject to both enzymatic and non-enzymatic deterioration. Fat or lipid changes include hydrolysis or breakdown into component parts, some of which may have off-flavors.

Probably the most visible alterations of beef are associated with color. The commonly accepted fresh beef color is a surface phenomenon which has been the subject of intensive research for many years. Details of the pigments and their reactions can be summarized into salient factors. The cherry-red color so common to retail meat cuts is oxymyoglobin, an oxygenated form of muscle pigment. This natural pigment is purple-red myoglobin, a color existing just below the surface of cut beef. At high oxygen levels, as when the beef is exposed to air, oxymyoglobin is formed. Upon removal of the air, the loose bonding of oxygen to the myoglobin leads to a reversal of reaction to form myoglobin. Myoglobin may be truly oxidized to metmyoglobin, a brown substance with iron at the ferric state. The latter reaction occurs most rapidly at relatively low oxygen concentrations, but occurs at a lower rate at high oxygen concentrations. Metmyoglobin formation is accelerated by factors which cause denaturation of the globin or protein element. These factors include heat, salts, ultraviolet light, and surface desiccation which causes increased salt concentration.

Metmyoglobin is a relatively stable compound, so the myoglobin to metmyoglobin reaction is not reversible without the presence of added reducing agents. This discussion clearly indicates no relationship between any deterioration factor and color.

The most important parameter affecting shelf life is microbial. Microorganisms are not inherently present in the tissues which are essentially sterile. Microorganisms are introduced from the air and from tools used to cut the meat. The ubiquitous presence of microorganisms virtually prevents their absence from the meat surfaces in any present-day commercial operation.

Microorganisms are living entities which grow and multiply, using the meat as their source of nutrients. The excretion products of microorganisms are the undesirable off-flavors of spoilage. As they increase in numbers, into the millions per gram of meat, their deteriorative effect naturally increases many fold with formation of acids and protein breakdown products.

Growth of microorganisms is accelerated by increasing temperatures up to their optimum after which growth rate generally decreases. At about 130-140°F, microorganisms are destroyed by heat denaturation.

In any nutritional environment such as meat, microorganisms go through a lag period at any temperature while they are getting established. Lag periods are longer at lower temperatures. The organisms then multiply in a logarithmic fashion up to about a maximum number the meat can sustain. The higher the temperature up to the optimum, the shorter the logarithmic period which is measurable in minutes or hours. The optimum temperature for microbial growth is approximately at room temperature. Thus, the most significant vector with a potential for reducing shelf life is also highly temperature dependent.

Generally, the microorganisms which contaminate meat are aerobic, requiring oxygen to grow. The metabolic processes involve an intake of oxygen and the production of carbon dioxide and water. Thus, reduction of oxygen and increase in carbon dioxide and moisture can retard microbial (as well as enzymatic) activity.

In addition, meat may undergo water loss which leads to economic losses as well as changes in color. Since meat is sold on a weight basis, loss of water by evaporation means loss of dollar value. Loss of water is a physical phenomenon in which vapor pressure of water within the meat, which is about 80% water, exceeds the vapor pressure of moisture in the surrounding atmosphere. The rate of escape of water from within the meat increases with increasing temperature. If the environment above the meat, however, contains its full complement of water vapor molecules, an equilibrium is established, and moisture loss from the meat is negligible.

All of the above points to four deteriorative vectors which can affect meat shelf life:

- Biochemical reactions
- Enzymatic reactions
- Microorganisms
- Dehydration[26][33][35][54]

**Fresh Meat—Packaging**

Packaging is a mechanism for meat to reach the consumer in a clean and usable condition. Basic requirements of packaging for meat products include prevention of contamination from external sources, such as dirt, dust, and microorganisms, in order to maintain as much of the initial quality as possible within the parameters of the external environmental conditions.

With fresh products, marketing requirements today dictate that the meat be red. To assure red color, oxygen is required by the pigments near the surface, and, thus, there must be oxygen within the package. Oxygen-permeable packages have been created for this purpose.

The film must be compatible with the meat; neither the material nor the tray should be punctured by the irregular shape of the meat; the package should conform somewhat to the irregular shape; and the material should seal so that it does not come off in transit, display or in the home.

Meat films now include both MSBO and one-side polyethylene-extrusion-coated cellophane, which is also oxygen-permeable, the polyethylene acting as a sealant and as a water vapor barrier. Cellophane is machinable and has been an effective meat wrapper.

As post-war technology progressed, newer plastics emerged. Although many materials were tried, only polyvinyl chloride (PVC) has gained major acceptance and has recently replaced cellophane. Less expensive than cellophane, although somewhat less machinable, it has a sparkling clarity and is considerably better in its ability to conform to shape, be punctureproof, and to return to its intial form after being handled by the customer.[1]

PVC can be tack sealed provided not too much heat is applied. It has good oxygen permeability and some moisture permeability, but not so much moisture permeability that great product shrinkage occurs. Stretch PVC film operates reasonably well on manual and semi-automatic equipment in back rooms of supermarkets. Being a stretch film, it has a memory so that when it is handled it returns to its original shape.

For automatic equipment in supermarket back rooms, biaxially-oriented shrink PVC film has been applied. When heat is applied, the film shrinks tightly around the package. Stretch films cannot operate on the semi-automatic equipment.

A successful example of vacuum handling of fresh meat product defies all the so-called rules of fresh meat operations. Chub packaged ground beef, however, has taken advantage of basic principles and, in its marketing areas, has consistently outsold conventional freshly prepared transparent packaged ground beef. The chub cylinder, made of oxygen-impermeable 50-75 gage, double wound PVDC, is filled on automatic form, fill and seal equipment. The package is filled so that there is no oxygen headspace; microorganisms inherently present and enzymes quickly use up all the residual oxygen, creating an environmental condition unsuitable for aerobic microbial growth. The meat color within this opaque package, however, is myoglobin purple-red, but this so-called undesirable color is not visible to the consumer. Further, package opacity excludes light which can catalyze irreversible color and fat reactions, and radiant heat from lighting which would penetrate transparent packages, raise the surface temperature of the meat and accelerate microbiological growth. The technical need for opaque packaging dictated total overall printing which has led to brand identification and consequent consumer recognition and an apparently higher return for the processor because of the higher price the uniform quality product commands.[4 19 94]

### Fresh Meat—Equipment

Machinery for packaging fresh meats has generally been designed for retail store backroom operations and, as such, cannot be considered very sophisticated equipment. The equipment generally consists of conventional overwrap machines with simple overlap heat-seal devices. They are most often fed, and they operate at speeds of 15 to 40 packages per minute.

Most back-room operations in retail supermarkets today use semi-automatic or manual packaging machines and stretch PVC or cellophane. Automatic equipment requires shrink PVC or cellophane, both of which are machinable. Equipment for fresh meat vacuum packaging is of two types: manual, in which a vacuum is mechanically drawn on a large cut of beef; and Kartridg Pak chub packaging machine, which uses double wound PVDC on an automatic vertical form, fill and clip machine. The latter operates at about 60 per minute.

The present conventional equipment has been made primarily by four companies for back-room operation where neither speed nor maximum labor saving is essential. Machines have been designed to insure a minimum of maintenance requirements. As one might expect, most of the wrapping equipment is an outgrowth of weighing equipment used to assure proper weights. Economy and legal requirements make precision weighing of meat products mandatory.

From the precise print-out scales has come ancillary, semi-automatic and automatic packaging equipment from Hobart Manufacturing Co., the Toledo Scale Corp., Franklin Electric Co., and Package Machinery Corporation. None of the automatic equipment has speeds greater than 40 packages per minute, which, for an overwrap machine, is slow. In addition, there is semi-automatic equipment available from these companies.

Unfortunately, no equipment adequate for the requirements of a centralized meat-packaging operation has yet been developed. Such equipment would have to be economical, require little maintenance, operate at low temperatures, and package different sizes of packages without roll size changeover or major operator adjustments.

Meat-packaging shrink tunnels are of the conventional type, and all tests have indicated that shrink tunnels have no particular deleterious effect upon the meat.

### Processed Meat—Packaging Requirements

Frankfurters and luncheon meats are subject to the same spoilage vectors as is red meat, with two important exceptions: additives aid in extending shelf life considerably and color has been stabilized by curing agents. Salt and other preservatives retard, but do not prevent, micro-biological spoilage. Removal of oxygen, how-ever, adds to the slowing of microbial pro-cesses. Color has been fixed by the addition of sodium nitrite and sodium nitrate to form a very stable nitrosomyoglobin pigment. For this reason, vacuum or reduced-oxygen packaging is employed.

## Processed Meat—Packaging

Oxygen-impermeable films require complete sealability by either heat sealing or clipping. Rigid packages or double-seamed cans, of course, are quite suitable, but such packages are expensive. The most widely used flexible film gas barrier is polyvinylidene chloride (PVDC) known more commonly as saran.

Most frankfurters are packaged in a two-film system, one being thermoformed, followed by product insertion. The second or flat film is af-fixed to the drawn film, the contents are subjected to vacuum, and the package is sealed. Before 1960, maximum draw was about 1⅛″ limited by the extensibility of the primary lam-ination, which at that time was a 75-gage poly-ester (commonly called Mylar®) with 2-mil polyethylene as the heat sealant. When meat packages began to be deep drawn to 1¾″ or more, an even more extensible film was need-ed, nylon 6 laminated to 2-mil polyethylene was introduced.

Some companies are using heavy gages of semi-rigid PVC to effect a gas barrier with PVC/PVDC/polyethylene as the flexible mate-rial. To reduce costs, some packers are using two different flexible films with a deep draw of nylon 6 laminated to PVDC, and the flat por-tion of 50-gage (0.5 mil) polyester, both with polyethylene as sealant.[4 36 48]

Probably the most interesting packaging sys-tem developed for processed meat has been the Oscar Mayer continuous system used for lun-cheon meats, frankfurters, and bacon. The Oscar Mayer frankfurter line begins with the conven-tional stuffing in casing made from regenerated cellulose. The frankfurters are linked, cooked, smoked, and stripped, and then packed in groups of five by vacuum pick-up devices. Meanwhile, at the head of the line are three double-head extruders, two for PVC and one for PVDC. PVC and PVDC resin are fed into the extruders. Films for top and bottom pack-

age webs are continually extruded into water and between rotating nips under water where lamination of the PVC/PVDC/PVC occurs. The three webs are sufficiently amorphous that lamination occurs cold.

A top and bottom web are made simulta-neously with the bottom web being vacuum formed cold upon leaving the water bath. Five frankfurters are released into the formed bot-tom and the second web is brought down over the franks. Vacuum is pulled from top, bottom and side. Vacuum on the top and bottom is re-leased after the pressure plates are brought down on the edges. The amorphous films seal under pressure without heat. The package is then heated in a chamber to crystallize the film and is die-cut to remove the excess film. At 120 five-packs per minute, there is a 15% film scrap rate.

Luncheon meat is automatically sliced to weight in piles which are manually placed on moving forms. At the head of the line, two ex-truders form a single web of PVC/PVDC/PVC into nips and a water bath. A web of semi-rigid heavy-gage yellow PVC extruded elsewhere is thermoformed and die-cut into the backing sheets. A coating of pressure-sensitive adhesive is placed on the backing and, simultaneously, a paper label is affixed. The meat slices are auto-matically dropped into place, and the top web is stretched over the meat and adheres by pres-sure to the adhesive ring.[54]

Bacon is sliced in a conventional manner and is manually weighed onto two-side polyethylene-coated, wax-impregnated paper. At a corner of the line are two extruders forming a top and bottom PVC/PVDC web. The bottom web is virtually flat but has a vacuum pulled beneath it. The bacon packages are automatically placed on the bottom web and the top web, vacuum pulled over the top, fuses with the bot-tom web at the periphery. A set of flying knives cuts the excess material.

## Processed Meat—Equipment

Vacuum packaging machines for processed meats are made by companies such as Standard Packaging Corporation or Mahaffay & Harder. These machines thermoform one sheet and seal a second sheet over the top filling. A vacuum is drawn just before final sealing. Both are widely used in high volume operations.

The gas-flush package has been primarily promoted by the Hayssen Company. It has had some success for a number of cured meat products. Gas-flush packaging is, according to Hayssen, actually "controlled sterile atmosphere packaging." It was originated several years ago in the cheese business and extended in about 1964 to the meat industry. Hayssen uses an "RT" machine redesigned for this purpose.

The system inserts cured meat product and substitutes carbon dioxide or nitrogen or a combination of carbon dioxide and nitrogen for the air, primarily to exclude oxygen. Gas-flush is not claimed to be a substitute for good refrigeration and sanitation. Carbon dioxide flush is equal to vacuum, according to Hayssen, in extending shelf life.

# POULTRY AND EGGS

## Summary

Poultry has emerged from a bleak past into a popular American food. Chicken, however, is generally handled like fresh meat, being cut and packaged in back rooms. Packaging is tray plus PVC overwrap. A new and rapidly growing concept has the bird in shrink packages, factory packaged and chilled to 29°F for extended shelf life.

Turkeys, because of seasonal demand, are generally frozen in cling film.

Eggs are still overwhelmingly packed and sold in the same way as two decades ago: molded pulp and paperboard cartons.

## Introduction

Partially as a result of increasing prices for red meat, poultry has enjoyed a healthy expansion in recent years. New merchandising methods based on the desirable flavor, appearance, and nutritional value of poultry have helped bring about the increase. Chicken represents one of the lowest cost animal proteins available to the American consumer.

Chicken is now being marketed in prepackaged form, cut into parts according to the desires of the market segment being served. Also, a very healthy business in frozen precooked chicken has developed.

Turkey, still peaking at Thanksgiving and year-end holidays, has become available on a year-round basis.

Poultry agriculture has become a business focused on bringing young birds to market. The killing and dressing industry has become a highly efficient operation. Sanitation standards of growing and cleaning, while not always ideal, have been vastly improved.

Although a number of advanced integrated operations exist, a large percentage of poultry is still brought to the retail market in water ice. The dressed birds are separated from the ice and cut into parts in the supermarket back room where packaging and weighing are performed.

Another similar program, which is growing, has the dressed birds packed in a corrugated container with solid carbon dioxide as a chilling agent in a separate compartment. This is cleaner and easier to handle than birds chilled in ice.

The objective after killing and dressing is to minimize microbiological and enzymatic degradation by rapid temperature reduction to as close to the freezing point as possible. Of all foods, poultry must be classed as among the most susceptible to microbial growth and spoilage.

In the past five years, prechilled, prepackaged poultry has become a most exciting development to the industry. Several companies have integrated back to the flocks and forward to the retail display cabinet. The birds are killed, dressed, cleaned, and packaged in a shrink film bag. Printed on this retail bag are the store brand, product description and guaranteed weight. After closure and shrinking to reduce headspace, the packaged chicken is blast chilled so that the surface freezes. Tempering then brings the entire body to 29°F, a temperature at which the poultry is maintained in storage, distribution and at retail. The company controls the product and so is able to guarantee quality and weight to the retailer.

Branded, quality merchandise is offered the consumer at a premium price which she evidently pays.

Similarly, frozen turkey in whole bird form has enabled the growers and processors to level their production. Thus, frozen turkey in cling bags has become a significant food purchase. It is relatively convenient to stuff and roast from the frozen state.

In contrast to chicken, eggs have declined in per capita consumption in the United States. Egg marketing or merchandising has not markedly improved since their entry into supermarkets. Preprepared eggs have not been introduced widely into the retail market. Dried eggs find their way into the retail market as ingredients of prepared mixes, cakes, and other foods. Frozen scrambled eggs have only recently been introduced into test markets.

The structure of some cooked egg foods is so critical that there is as yet no really good way to preserve textural characteristics. Omelet textures have been frozen for several years in ethnic dishes such as egg foo yung, but scrambled and cooked egg structures are not retained by freezing.

Many of the eating properties of eggs reside in their structure. The albumin and yolk of the egg can be separated and held under refrigeration for some time, or even frozen.

Shell eggs are handled in much the same manner today as they were twenty years ago. Laying flocks are perhaps larger, and some gathering methods are more efficient. Large, fully integrated lines are used to package cleaned shell eggs.

But the eggs are still distributed in molded pulp and complex paperboard cartons under minimum refrigeration. Shell eggs are sold unbranded by size and shell color, with retailers generally attaching their own imprint to the package. Shell eggs are packaged at central locations, some of which might be regarded as regional centers. Depending on market demands, they are sometimes stored under refrigeration for several months. Eggs do lose flavor on storage and, of course, are quite susceptible to contamination from outside odor sources. They are indeed fragile, but shell eggs are hardly as susceptible to bacteriological deterioration as are chicken and ground meat.

Despite the many proposals for innovative packaging, there has been virtually no followthrough. A small quantity of shell eggs is packaged in a tray overwrapped with a shrink film to display the contents. This package represents flexible packaging's only commercial entry into shell egg packaging.

Shell eggs are breakable, inconvenient to handle, somewhat difficult to work with in the home, and not always attractive. The edible contents are not sufficiently delicate that they could forever defy imaginative packaging and marketing.

## Packaging Requirements:

Poultry is an animal muscle tissue subject to enzymatic and biochemical breakdown. The tissue is inherently sterile, but microorganisms can contaminate the surfaces. Removal of the intestinal cavity contents often leads to spreading of microorganisms although in new dressing plants, the problem has been considerably alleviated. Still there is no practical mechanism for assuring the absence of microorganisms.

Good refrigeration retards microbial as well as enzymatic and biochemical activity. Prevention of added contamination is another requirement since the microbial load has a direct effect on shelf life.

Moisture loss should be retarded since poultry is sold on a weight basis. Low temperatures reduce the rate of moisture escape, but packaging is a key to prevention of dehydration and attendant weight loss and loss of appearance and texture.

The fat is subject to oxidation, but this does not seem to be of concern to packagers, because this defect follows other degradation.

Shell eggs are inherently sterile unless the integrity of the shell has been broken. Egg contents can lose some of their desirable structural properties as a result of enzymatic or biochemical deterioration which are a function of temperature. The rate of these changes is reduced by temperature reduction, but they are still relatively slow, i.e., on the order of days at ambient conditions.

The most important packaging requirement for shell eggs is retention of shell integrity. For this reason, elaborate cushioning devices have been designed and used to protect eggs against shock and compressive stresses.

## Packaging

Because poultry is generally handled in much the same manner as meat, it employs the same packaging components. Retail stores cannot afford totally different systems for red meat and poultry packaging. Therefore, although their requirements are somewhat different (poultry does not need oxygen), they are packaged in the same way.

Back-room poultry packaging involves the use of standard size trays. After manual filling, the trays are overwrapped with PVC film which retards added contamination and moisture loss. While providing transparency, PVC film also is tough enough to withstand some sharp and irregular shapes.

In-plant packaging uses a totally different concept in packaging. Content degradation is retarded by two mechanisms, low temperatures and effective exclusion of oxygen. The latter retards the propagation of aerobic microorganisms. To protect the contents, a cling film system is employed, e.g., preformed Cryovac PVDC bags into which the bird is packed. The package is clip closed, and the film is shruck around the bird by application of heat. This method is effective, protective and attractive.

The same system is employed for frozen turkeys. Shrinking the film effectively excludes oxygen.[28][51]

Shell eggs in flexible film use PVC film over pulp, paperboard or foamed polystyrene trays, each with a conventional 6 x 2 cell arrangement.

### Packaging Equipment

Poultry packaging uses exactly the same machinery as red meat packaging. Most is manual wrapping with roll unwinds and hot plates for heat tacking. Typical equipment is by Hobart or Cleveland-Detroit. Hobart has also adapted the Corley-Miller machine for semi-automatic tray wrapping.

A number of back-room operations use automatic equipment such as the Franklin or Dove and the Package Machinery Corp., U-6.

Cryovac packaging equipment is supplied by Cryovac Division of W.R. Grace & Co.

# PRODUCE

### Summary

Produce is a group of commodities which is seasonal, perishable and the victim of phenomenally high waste because of incredibly poor handling practices. Only a few items are prepackaged, and many are not packaged even in back rooms. Produce is without question the least rationalized of all food industries.

### Introduction

Perhaps the most non-rationalized portion among all American food industries is produce. To arrive at the consumer level, fresh fruit and vegetables proceed through some of the most complex channels modern society has produced. Partially as a result of the limitation of these channels, consumer dissatisfaction with the results has been reflected in a downward sales trend for these items.

The local wholesale market still exists and serves as a distribution center for produce grown locally and at distant points. With few exceptions, preservation techniques are unknown to this industry, and so seasonality is a major factor. The agglomeration of agriculture has led to improvements in yields and appearance, and speed of production, but eating quality has somehow not received as much attention.

Each fruit and vegetable variety in a given area has a brief harvest period during which it may be picked, shipped and eaten. During periods in which there is no domestic crop, produce may be imported if there is a real demand. A few products, such as bananas, are imported throughout the year because they are not grown commercially in the United States. Tomatoes are grown under artificial conditions or transported great distances to supply the year-round demand.

Only a few fruits and vegetables are in sufficient demand to be found at retail throughout the year; lettuce, tomatoes, potatoes, celery and bananas are examples. Others, such as strawberries, blueberries, peas, eggplant and cauliflower, are available only for brief times because of their harvest periods. Grapefruit and melons are available for most but not all of the year by purchasing from a constantly changing series of growing regions.

The supply situation leads to a plentitude during harvest period and pressures to move produce as rapidly as possible. Only a few products, such as potatoes and some apples, are stored for later sale.

To move produce from source of supply to the consumer under these circumstances are brokers, jobbers, wholesalers, direct buyers, and any number of middlemen and transportation mechanisms. Produce often passes through many hands before arriving at the retail level.

The loss rate as a result of multiple handling and lack of preservation methods is frightful. Still, retailers do not elect to drop produce because it is an accommodation to the consumer. Introduction of the most elementary of preservation methods would go a long way toward improving the situation, but there are so many different handlers involved that no single one apparently wishes to initiate a system. Only those products which would literally spoil overnight are preserved.

The principles of good preservation are well-known and might be applied. Low temperature just above the freezing point sufficiently retards enzymatic respiration of most fruits and vegetables so that they can be distributed to the consumer with good quality. Produce native to warmer climates may be damaged by 32°F temperatures, but each has a below ambient temperature optimum for its shelf life extension. Among the products susceptible to this type of damage are avocados, bananas, eggplant, grapefruit, peppers, pineapples, potatoes, and green tomatoes.

On the other hand, ambient temperature handling accelerates breakdowns and microbiological deterioration. Ambient temperatures also accelerate loss of water, leading to weight loss and dehydration, wilting and similar defects.

Therefore, for maximum preservation, produce should be reduced in temperature to the lowest point above which chilling or freeze damage does not occur. It should then be packaged to retard moisture loss, and should be maintained under these conditions until retailed.

Unfortunately, these practices do not prevail. Some produce is chilled after harvest to remove field heat. Lettuce, for example, is often vacuum cooled. Sweet corn may be hydrocooled or sprayed with chilled water. Apples may be forced air cooled. Most produce, however, undergoes only the minimum cooling processes. The longer the distribution channel for which the produce is scheduled, the higher the probability that there will be some chilling. When one considers the number of handlings and

transfers the produce goes through, however, the probability of adequate refrigeration throughout produce life is ly reduced.

Little produce is packaged near the growing area because the growers and wholesalers do not always know the final destination. Rather, most produce is bulk packed in a broad variety of containers. At some point near the retailer operation, the produce may be consumer packaged.

A few products like radishes, oranges and potatoes, all of which are hardy, may be packaged by a grower or a grower's cooperative. In a few areas, some packers package and supply a range of produce under a brand name. Because the produce can spoil rapidly without proper handling, the brander may have an undue amount of defective goods due in no way to his own negligence or the produce quality.

A result of this multiple handling and seasonality is considerable unpackaged produce at the retail level and some back-room packaging.

The spoilage rate is extraordinarily high because of lack of adequate refrigeration, stock turnover, attention to initial quality and general consumer disinterest. The consequence of these practices is that produce is almost totally a commodity. With little margin for profit, few in the distribution cycle want to invest to improve the situation.

A small number of advances have been made. The very use of any prepackaging represents a step forward. Some entrepreneurs are beginning to cut and trim vegetables to consumer sizes and shapes and packaging them, as for example, precut green beans and celery. Prepackaged lettuce is not uncommon. Bananas are often branded, if not consumer packaged. Apples out of storage may be prepackaged.

For purposes of this discussion, produce may be divided into four categories:

1. Vegetables for fresh consumption: carrots, celery, lettuce, onions, tomatoes, cucumbers, green onions, and radishes.

2. Vegetables for cooking: corn, onions, potatoes, peas, sweet potatoes, beans, cauliflower, squash, asparagus, beets, eggplant, turnips, broccoli, spinach.

3. Short shelf life fruits: bananas, peaches, watermelon, grapes, pears, plums, cherries, strawberries, and blueberries.

4. Longer shelf life fruits: apples, grape-

fruit, oranges, lemons, tangerines, melons and pineapple.

## Packaging Requirements

An assumption of proper handling so that initial quality is present and preservable by packaging would be erroneous. The situation described could be altered somewhat by appropriate packaging, but would require a drastic upheaval to bring order and high quality to the marketplace. To apply excellent packaging to produce when it is normally packaged is almost absurd. Packaging cannot improve, it can only help to retain what is present. Good packaging can only be applied when it is accompanied by good handling and distribution.

The requirement for primary packaging material at present, then, is to prevent further damage from the packaging procedure. Consider only that backroom prepackaging occurs after the produce has survived most of its usable life.

All fruits and vegetables are respiring entities, employing atmospheric oxygen, and expelling carbon dioxide and water. Respiratory action can be reduced by lowering the temperature. It may also be reduced by lowering the oxygen concentration and/or raising the carbon dioxide and water concentrations. If, however, the oxygen concentration drops too low, anaerobic respiration or fermentation can occur. Excesses of water vapor in the surrounding atmosphere can lead to condensation on the produce surface and stimulation of mold growth.

Because of the hazards of anaerobic respiration under normal handling conditions, packaging requirements today stipulate that oxygen and carbon dioxide freely pass. Further, because temperatures are generally not low enough to retard mold growth, packaging requirements should allow some water vapor to leave the immediate environment of the produce. A slight drying is generally encouraged; and, thus, an added requirement is that the packaging, if transparent, does not fog.[58]

Transparency is a marketing or merchandising desire, sometimes observed in practice, and sometimes not. Generally, there is a desire for at least partial transparency so that the consumer can see the produce.

Fruits and vegetables are odd shapes and sizes, so packaging should contain these forms without damage, but with some semblance of holding them in place.

Because so many fruits and vegetables are tender and subject to physical damage if shocked, there should be a requirement for structural protection. Unfortunately, this ideal is observed more in the breach.[5][19]

## Packaging

Retail back-room packaging has a common denominator of tray plus film. Generally this is the same type tray plus basic film used for meat and poultry packaged in the store. Packaging not performed in the store is usually specialized for the produce involved, but not sophisticated enough to improve preservation.

Thus, backroom packaging involves the use of PVC film of produce grade, that is, with a slight green tinge to show off the product to best advantage.

Some retailers use trays such as for meat, and some purchase special trays colored for produce. Some retailers tray all produce before PVC packaging and some tray only a few of their items, thus saving on the cost of trays.

Packaging in other than the backroom is a variety of inexpensive forms. Apples after removal from storage may be packaged in polyethylene bags, often perforated to assure against fermentation, excess moisture and mold growth. (Apples are very often distributed and displayed without refrigeration).

During distribution, hands of bananas are bundled in polyethylene sleeves that hold the fingers together and help protect the bananas from physical damage.

Carrots may be bunched into polyethylene bags. Again, perforations may be used to assure against possible respiratory damage as a result of poor temperature control.

Celery bunches are frequently partially packaged at the base, part way up the stalk. Generally, parchment or polyethylene is used to hold the stalks together rather than for any real protective purpose.

Grapefruit are sometimes bagged using perforated polyethylene, more as a means of holding together a number of fruit than for any protection. More recently, polyethylene netting has been used for this purpose.

Lettuce is often prepackaged at various sites between cooling and the retail store. Polystyrene has been found to have appropriate permeability characteristics and has found wide use for this purpose. Some perforated polyethylene and some PVC's are also used. The film's principal purpose is to hold the leaves tight to the head and retard wilting of exterior leaves. This helps prevent gross losses.

Dry onions are packaged in perforated polyethylene or cloth net bags. This produce is relatively stable, but can be damaged in very long-term storage or when wetted.

Oranges are generally bagged only as a means of carrying them. Thus, cloth net, perforated polyethylene or polyethylene net bags may be employed.

Potatoes are most often packaged in windowed multi-wall paper bags. A small number are in perforated polyethylene.

Tomatoes are generally trayed or cartoned because they are very soft. Injection molded polystyrene frame trays, after filling, are overwrapped with polystyrene or PVC film. An opening is often left to assure against fermentation. Both films may be at least partially shrunk to attempt to hold the fruit without jostling damage. Paperboard cartons for tomatoes have polystyrene windows.

Some grapes are now packaged in polyethylene netting.

Radishes are the only produce packaged in impermeable packages. Generally polyethylene is used; a controlled atmosphere is thus created within the package.

Packaged mixed salad vegetables or cole slaw cabbage are often in a three-side sealed polyethylene pouch. These products have a very short shelf life and so are delivered on a daily basis.[45]

## Packaging Equipment

Because there is so little packaging except in the back room, there is little produce packaging equipment. Backroom machinery is the same as that used for meat.

Most prepackaged produce is manually packaged with shrink tunnels being the only equipment used. Some bagging equipment exists, but this is mostly for preformed bags.

Several U.S. Department of Agriculture studies have indicated savings in materials, labor and waste by removal of produce packaging from the backroom to central locations.

Some chains are purchasing centrally packaged produce, but the practice is not widespread. The strong belief that produce is a differentiating factor between stores has held up much movement of produce packaging away from stores.[34] [55]

# SNACKS

## Summary

Snacks represent a large and growing food product area. It is the subject of intensive packaging, product and marketing research and innovation. Led by potato chips, which continue to grow despite assaults from other products and from other food areas, snacks have continued to undergo a series of packaging changes.

Potato chips, because of their volume and perishability, are generally of local or regional manufacture. They, and most of their successor products, are delivered and stocked by manufacturer route salesmen. Attempts to upset this expensive distribution system with long shelf life snacks have not interrupted the growth of potato and corn chips. Assurance of shelf space and proper stock rotation appear to be valuable to the manufacturers.

Potato and corn chips are packaged in three size classes, small, medium, and large. All are in vertical form, fill and seal pouches. The small are generally coated glassine; medium in glassine/polypropylene or cellophane/polypropylene and large in waxed sulfites.

Long shelf life products are foil laminated bag-in-box or DPM systems or composite cans.

## Introduction

Snacks are the glamour issue of the food industries. Snack foods have been the recipient of boosts from many directions: new manufacturers, suppliers, marketing, distribution, and packaging.

As people have trended away from the conventional three-meal day because of lack of time or because of an excess of time, they have begun consuming many mini-meals. Snacking is sometimes referred to as the continuous fourth or fifth meal. Even those who adhere to the three meal a day concept consume snacks.

Potato chips now constitute the largest single product category of the snack food industry.

# FIGURE 3

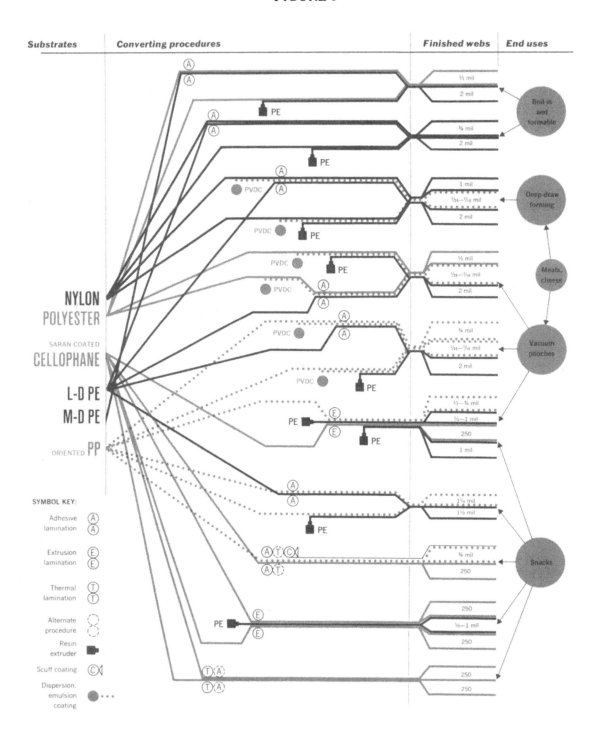

Examples of converting procedures employed with several laminating materials to produce flexible packaging structures for food packaging.

Reprinted from *Modern Packaging*, October 1967. With permission.

But potato chips now include flavored and formulated potato chips, and chips in a variety of shapes. Far behind, but growing more rapidly, are corn chips, now in flavored and tortilla versions. Pretzels, pretzel sticks, puffs, french fried onion rings, and chips fabricated from other starchy materials are also included.

On the periphery are a group of crackers specifically designed and marketed as snacks. A few meat products such as bacon rinds, jerky and dried sausage have enjoyed a rebirth. Several ready-to-eat cereals have partly fallen into the snack category.

The venerable snack has been a chip, fried in oil, bag packaged and eaten as a between-meal item. Potato chips are produced by slicing peeled potatoes and deep fat frying. Because of the fat content subject to rancidity when spread over such a large surface area, distribution and turnover have had to be rapid. Despite intensive study, there has been no dramatic commercial improvement in shelf life stability of potato chips. They remain a product requiring special distribution channels.

Corn chips are made from a prepared dough which is extruded, cut and fried. Pretzels are baked after formation from dough. Puffs are expanded during extrusion.

Formulated chips are made from ingredients which include potato mixed into a batter or dough and dried or baked into chip shape. Frying is used to impart flavor and mouthfeel, but the amount of fat is considerably lower than for conventional chips. The manufacturers distribute these products as conventional dry groceries with long shelf lives.

More recently, a new class of snack item based on cereal technology burst on the market. The flour dough here is extruded and toasted. Flavors are incorporated in the mix and on the surface. Again, the marketing is as long shelf life products.

The success of potato chips and then corn chips spurred the imitators and the innovators. Apparent defects in potato chips, such as tendency to rancidity, occasional sogginess and fragility, were attacked by the newer products. Still, potato chips grew, and continue to grow. The rate of growth is only steady, slightly above that of the economy.

Because of space volume and perishability, potato chips are regional and local products produced by smaller firms or by regional factories of multi-plant firms. They are almost all delivered on a several times a week basis by manufacturer route-salesmen who stock the shelves and rotate the merchandise. Potato chips have a very low density, and thus occupy enormous amounts of shelf space. To attempt to stock more than a few days' supply in any retail outlet would mean occupying a disproportionate amount of shelf space.

Potato chips are sold in several packaging forms: the small loose pack selling for five or ten cents; multipacked small packages; medium-size packages, the largest seller; and large or twin size packages, a relatively small factor.

Potato chips are generally programmed for a three-week shelf life. The importance of the route salesman thus becomes apparent.

While the potato chip industry consists of a large number of firms, the corn chip industry is dominated by one firm. Package sizes are as with potato chips. Almost all corn chip manufacturers make and sell potato chips, so the distribution system is the same.

Pretzels tend to be regional items, selling extremely well in places like Pennsylvania, but doing well in other areas. Thus, manufacturing tends to be concentrated in areas where they sell well. They are generally made by companies that manufacture only pretzels. Distribution, however, is by route salesmen, with the cycle being longer because pretzels are more stable and dense than potato chips. Further, turnover is considerably slower.

Puffs tend to parallel pretzels: puff manufacturers specialize in puffs.

The newer formulated chips and cereal-based products are packaged to minimize deterioration. Distribution is as conventional grocery items through retail store warehouses. Shelf life is programmed for six to nine months. This product is more dense and less fragile than potato chips.

While each category of long shelf life snacks enjoyed an initial spectacular growth, each has fallen off considerably. Formulated chips rose and then declined; cereal-types were thought to be a success but have at best leveled. Still, the products continue to be introduced. Formerly in paperboard cartons, the latest wave is in composite cans.

The biscuit and cracker industries have expanded into snacks by introducing crackers with snack characteristics, and by purchasing true snack food firms. Cereal companies have also taken both paths.

The demand is not yet satiated, and no end appears in sight. It is difficult to project what the next step will be. For many years, there have been predictions of the end of the expensive route-salesman system. But the system serves to assure shelf space for fresh product. Regular delivery is a necessity, anyway, and the chippers have the distribution system. While expensive, the route-salesman system still results in profitable operations for chippers.

It is easy to condemn a system and to offer what appear to be logical improvements. These have been tested in the fire of the marketplace, and the existing system has continued to flourish. The newer snacks have found niches, but the major effect may have been to expand the sales of all snacks.

Of all food product categories, snacks are without question the most dynamic and unpredictable. Only potato chips are steady, and that only in the basic product. All other factors about them change. Today's packaging and marketing may be obsoleted by press time. There has been an historic precedent of change.

## Packaging Requirements

The key to packaging requirements for all snacks, and particularly among the deep fat fried, is the presence of edible oil in large quantities. The oil is spread over enormous surface areas and thus exposed to the environment.

Unsaturated fatty acids in fats are subject to oxidative rancidity in the presence of air. These reactions are accelerated with temperature increases, moisture, by light, and by salt. The last named is a usual constituent of a snack flavoring system. Antioxidants help, but no antioxidant can block all oxidative rancidity reactions in a deep fat fried product like potato chips.

Although there has been much experimentation and some commercialization, generally potato chip packaging has avoided exclusion of oxygen. The belief is that deliveries must be made to keep the shelves filled and, while oxygen-free packaging would extend shelf life, it is an unnecessary expense. Some concession to

the problems of oxidative rancidity is made by the use of antioxidants.[27]

The presence of fat indicates that a major requirement of snack packaging is greaseproofness. This requirement is important to reduce the surface area of oil and thus retard rancidity, to prevent unsightly staining of the package and to prevent actual seepage and greasy package feel. Oil can also in some instances lead to splitting of lamination by dissolving the laminant.

One of the major characteristics of snacks is crispness. Crispness is achieved by creating a structure and then drying it by frying or baking. Retention of the desirable texture is directly related to water content. The water content of snacks is very low; any increase, because of the high surface to volume ratio, can lead to a loss of desired crispness. Further, added water also accelerates other biochemical changes such as oxidative rancidity. Low water vapor permeability is, therefore, another very critical packaging parameter.

Packaging should act as an odor barrier to assure against contamination of the product and to assure against contaminating other products. Light is a rancidity catalyst but there is considerable debate from marketing standpoints as to the value of opacity for technical protection versus transparency for merchandising. The situation has hardly been resolved, with parts of the country such as the East and Southeast mainly in transparent and the remainder of the country being, for the most part, in opaque packaging.

Machinability is a major packaging requirement. Snacks have only recently advanced away from manual filling of preformed bags into form, fill and seal equipment. As a result of the late entry, the snack industry is using late model high-speed machinery. Packaging materials must, therefore, be capable of operating continuously and efficiently on some fairly sophisticated equipment.

The question of stiffness is also debated between technical and marketing men. Marketers want materials that can stand up on the shelf, but greater stiffness leads to machinability problems.

Much has been written and spoken of cold weather durability, and because of the dependency on flexible packaging with a minimum use of true plastic and heavy use of hydrated cellulosics, cold weather durability may be a factor. There are, however, other food industries with apparently the same problems who have not found the problem so serious.

The fragility of most snack items would indicate that the packaging material should provide some structural protection, but because of direct manufacturer control, there is apparently a belief that such protection is not required. Long shelf life snacks do not enjoy benefit of gentle handling, and so require structural protection beyond that of a flexible package.

### Packaging

The overwhelming majority of snacks today are in flexible bags. The use of the route salesman system has enabled the snack foods industry to use less expensive packaging than would be employed if warehouse distribution were employed.

Snack food flexible packaging, i.e., the conventional 21-day shelf life potato chips, used preformed heat sealing glassine and cellophane bags until just a few years ago.

Single wall cellophane gave way to duplex because of the bag breakage problem. For added moisture protection, polymer coated cellophane was used when it became available. Again, the use of duplexing in medium and larger bags was common, and this was an expensive combination. It was neither cheap to purchase nor cheap to run on early vertical form, fill and seal machines used for chip packaging. The chippers, mostly small, did not seek, but they hoped for a single web that would satisfy their need for an inexpensive sheet.[20]

Polyethylene/cellophane was satisfactory as a barrier but required a fin back seal, meaning that existing equipment would have to be modified and that added material would be required for the additional seal area. Cellophane laminated to cellophane was used and sealed satisfactorily, but the laminating agents left much to be desired.

Oriented polypropylene had good weather durability, but was difficult to seal without shrinking the film. Coating the polypropylene with PVDC made fair seals possible, but the film is a relatively soft one with no stand-up character. Oriented polypropylene laminated to polymer-coated cellophane was successful for some time, especially those in which an overlap back seal was made possible by having one laminant shorter than the other. Opaque polypropylene has only recently been introduced. Meanwhile cellophane/polypropylene continues to be used for medium-size bags in marketing areas in which transparency is desired.[46][59]

In geographic regions demanding opacity, however, this was not a suitable lamination. Glassine laminated to oriented polypropylene is now the lamination of choice for medium-size bags. The glassine is mostly one-side PVDC coated for moisture and grease protection; two-side has recently been introduced. Generally the polypropylene is inside and the surface printed glassine outside with the PVDC coating towards the product. A number of methods are used to assure an overlap heat seal including a printed thermal stripe and an ethylene/vinyl acetate copolymer overlacquer on the glassine face. The laminating agent may be adhesive or PVDC or some other scheme.

This lamination is today considered satisfactory for chip packaging. The glassine serves as a fat barrier in case of any pinholes or defects in the polypropylene or the PVDC. Further, use of glassine retards passage of fat through the cut edge of an overlap seal. The various PVDC and polypropylenes serve as moisture barriers. The sheet is stiff, opaque and machines well on vertical form, fill and seal equipment.

Some chippers prefer a version in which the polypropylene is on the outside. In this lamination, the polypropylene may be reverse printed or the glassine may be surface printed. Although there is better surface gloss with this lamination, there may be less cold weather durability.

The West Coast has a number of laminations, one of which is polymer coated cellophane, reverse printed, and laminated to PVDC coated glassine. As might be expected this sheet is said to not have the cold weather durability of the polypropylene/glassine structure. Aluminum foil laminations are also appearing on the West Coast as well as in Canada.

While great attention has been paid the medium-size bags which constitute well over half the packaging material used, the small bag has very much stayed with one or another version of glassine. Various coated single webs of glassine or glassine wax laminated to glassine have been used for the nickel and dime bags. The coatings are PVDC plus ethylene/vinyl acetate copolymer or similar heat sealant. Smaller bags are probably going out of wax laminated glassine and into coated single sheets. As larger chippers put more small bags into overwrapped multipacks, there is less need for protection in the immediate wrap.

Very large bags and twin packs make use of some web stock and some preformed bags. Much of both is waxed sulfite although glassine may be used for the inner bag. A jumble of inexpensive combinations involving waxes and various papers is used for large bags.

Corn chips generally appear in the same types of packaging as do potato chips, except that they are almost all in opaque webs. The newer tortilla chips are transparent combinations to show the product.

Puffs are in a variety of flexible film form, fill and seal pouches, often polyethylene even though greaseproofness is an important characteristic. Pretzels are rarely bagged. Rather they are nested if they are large, or perhaps dump-filled in glassine-lined paperboard cartons if they are small.

Long shelf life snacks first appeared in bag-in-box systems. The inner bag is a paper/aluminum foil/polyethylene, three-side pouch style. Pouches may have been formed outside of the box and then inserted into the carton after filling and sealing. Pouches may be part of a bag-in-box system attached to the sleeve during carton manufacture. Also used frequently is the double package maker system using glassine/wax/aluminum foil/wax/glassine or paper/polyethylene/foil lamination. This inner pouch is dead folded to effect the closure and obtain protection.[20]

Proposals for oxygen exclusion or gas flush packaging have not been widely accepted. A few isolated chippers have tried it. To assure that no oxygen will enter over the long shelf time, a barrier such as aluminum foil must be included in the lamination. Most present structures do not satisfy this requirement.

**Packaging Equipment**

The need for a chip pouch to be voluminous and not constrict the contents is obvious from the fragility of the contents. This restriction has precluded the possiblity of four-side seal envelope type pouches.

Gusseted and satchel type packages were commonly used to assure sufficient space for the fragile chips. Such pouches could not be made efficiently from web stock on form, fill and seal equipment until recently. Conventional vertical form, fill and seal equipment made pouches, but had long narrow drops from the product reservoir into the formed pouch. The contents are dry and thus valuable. Automatic pouching of chips required gravimetric filling and short gentle drops into the pouch.

The Woodman machine is a slanted vertical machine so that the chips slide down an incline into the three-side seal pouch. In the even more popular Mira Pak, there is no tube. The forming collar feeds directly from the gravimetric filler into the formed pouch with a resultant short drop for the chips. These two vertical form, fill and seal machines, designed for and marketed to the chip industry, have between them almost all automatic chip packaging. They represent a philosophy of seeking out a market need and solving the problem commercially.

Long shelf life snacks have been packaged on double package makers, on Interstate bag-in-box systems, and by using conventional vertical form, fill and seal machinery and mechanical or manual insertion in paperboard sleeve.

## PRODUCTS/MATERIAL/MACHINE INTERACTIONS

A package must perform almost all of several functions: product protection, distribution, transportation, stacking, selling, communication of contents, opening, dispensing, reclosure, but, above all, containment. The consumer buys and uses the product and not the package. Packaging is a total system, not just the physical film material surrounding the product, but rather the equipment, people, designer, marketer and retailer who are involved, and the consumer who uses it.

In package development, the product has generally been defined in advance, i.e., the product entering the package is the product which must ultimately be removed. The term package refers not just to the primary material immediately surrounding the product, but also to the container in which the primary package is placed, labeling, coding, corrugated casing, etc.

All of the individual package elements must fit together. The primary package must fit not only product requirements but also the demands of the secondary package which, in turn, fits the needs of the distribution channels. The primary package must fit the demands of the retailer as well as those of the consumer who is ultimately to use the product.

Fundamental to the development of packages or products is the interaction of packaging materials, not only with other packages and with product being contained, but with the machinery which fabricates the package. In today's economy, manual operations are often used; but most high volume, high velocity foods are packed on semi- or fully automatic machinery. Thus, interactions exist between primary and secondary packaging materials and machinery used for the packaging operation.

Packages may be formed, filled and closed entirely by hand. Primary packages may then be enclosed manually in the secondary package. Even here, the materials must meet the singular demands of hand packaging.

Although much packaging is by hand, the large volume is performed with electromechanical assist, even automatically. But most food packaging operations include some people to provide for inspection or because it has been difficult to design machinery to perform adequately. Further, many operations do not generate sufficient production volume to warrant purchase or design of automatic or semi-automatic equipment. The number of different sizes, irregularity of shapes, short seasons, etc. have all served to discourage automatic or semi-automatic equipment for many food operations.

Two basic types of equipment are commonly designed and installed. One involves utilization of a preformed package which is filled and closed. The other involves enclosing the product in the packaging material and closing. In the first instance, the package has been preformed at another location and then brought together with the product. In the second, the package is formed using the product as a form or mandrel. Preforming may occur either in the packager's plant or in the converter's manufacturing operation. Preforming may be a segment of the packager's equipment cluster, coupled with the packaging operation; or it may be accomplished at one location and the package moved to the packaging operation. Bread bagging, for example, generally employs bags preformed at a converter's plant. Form, fill and seal equipment, in contrast, works from roll stock and forms the package around the product.

## Bags

Bags are generally loose fitting flexible containers which are preformed, filled with a product, and sealed or closed to retain and protect a product. Materials used for bags include: polyethylene; paper; cellophane; and, recently, some polypropylene. Because bags are often used to contain relatively low-cost products, decoration is generally inexpensive. For example, flexography is frequently employed to decorate polyethylene bags; central impression presses help reduce the distortion of the printing due to film stretching. Rotogravure printing is occasionally used for polyethylene. Paper used for high quantity bags is rotogravure, letterpress, or even offset printed. Low-cost printing processes are used for preformed bags, especially for low-quantity lots.

Polyethylene bags with salable transparency can be made for merchandising products such as bread. They can be automatically formed from roll stock by weld sealing on, for example, Schjeldahl equipment, to make a relatively inexpensive type of package. Polyethylene bags may also be formed from tube stock with a simple weld seal at one end so that the bag may be filled from the open end.

The major large-scale use of preformed bags on automatic equipment is for bread. Relatively few bread bags are formed in line with packaging.

Plastic bags may adhere together because of static electricity build-up and, consequently, may be difficult to open. Air blast is used to open the bags, but excess air leads to bag bursting. A hole in one end helps keep the bag from bursting.

Polyethylene can stretch and stick to metal causing machineability problems. Polyethylene and related soft films have virtually no dead-fold characteristics, i.e., they are limp materials. Some plastic films tear relatively easily but the tear is non-propagating. In contrast, cellophane tears easily, and the tear can propagate rapidly.

Preformed bags are usually used where the volume of production does not warrant automatic equipment. Preformed bags are employed for products not adaptable to direct in-line form, fill and seal operations; where low-speed operations do not merit a fully automatic equipment; or where packaging equipment must be moved from one place to another.

Bags are generally used for relatively specialized products which are not mass produced and for fragile or soft goods which automatic equipment cannot accommodate.

A common method of packaging in polyethylene bags employs semi-automatic equipment in which the bags are stacked on wickets or posts. An air blast opens the bag and the product is manually or semi-automatically filled into the open end. Pushing the filled bag tears it off the wicket. The bag falls to a station where it is simultaneously sealed and cut off with the generation of scrap polyethylene. Hamburger roll baggers such as the Pneumatic Scale equipment are of this type, with a clamp or reclosable tie seal used to bunch the end and close it.

Soft goods and other difficult-to-handle products are often packaged by forming a pillow-package around a product. For this purpose the polyethylene or polyolefin is formed into a roll of folded stock. In this way a two-side heat sealer in the form of an "L" can be used. The product is inserted between the folds and the "L" unit seals and cuts two sides. The "L" sealer thus forms two sides of the pouch with one edge simultaneously forming the edge for two different pouches. The contained product must be able to withstand external forces because the seal is often insufficient to take advantage of the protective characteristics of the materials.

In recent years, polypropylene and other polyolefins have been replacing polyethylene to some degree. Polypropylene is stiffer and more transparent than other polyolefins.

At one time, most preformed bags were formed from cellophane. Cellophane was the first of the major flexible films; but after many years of development and use, it still suffers from the problems of aging and brittleness and relatively little protective ability when compared against other plastic films. Cellophane's protective properties are in the coatings. Being stiff, cellophane is easier to print and is more machineable than other films which tend to stretch on presses and packaging machines. Cellophane, therefore, remains among the most widely used flexible films. Cellophane's growth has been arrested, however, because of deficiencies in performance.

Bags from cellophane may be gusseted or not depending on the contents. Cellophane bags can be easily slipped off a stack. With coatings, cellophane is relatively easy to heat seal. Because the base is cellulose, however, if heat sealing is performed at too high a temperature, burning could occur. With plastic films, heat could cause excess melting in the seal area. It is important that cellophane's coating be anchored to the base stock to assure that the bag and not just the coating is sealed.

In printed bags (or any other type of packaging) the inks must not be in contact with the food product contained. There are no known commercial inks which can be economically used in contact with the food product, either from a safety or aesthetic standpoint. Some vegetable inks may be used in contact with food product, but these could have a tendency to be soluble in food and cause color changes in the product. Thus, in an unlaminated film, the ink must be carefully adhered to the surface so that the surface printing does not scratch off during packaging or subsequent distribution. When laminated stock is used, the ink is usually locked between two layers of film and thus cannot come in contact with the product except when printing is between cellophane and polyethylene, in which case the sealing heat could conceivably release ink into the product.

A large number of machines such as the

*Doughboy* and Telesonic exist for filling and sealing preformed bags. On the other hand, the AMF Mark 50 machine is the most widely used for filling and closing bread. FMC Corp. makes equipment to form and fill bags from roll stock.

## Flexible Form, Fill and Seal

Flexible form, fill and seal packages are pouches formed simultaneously while packaging the product.

Automatic equipment for production of flexible packages is basically of two types: vertical form, fill and seal; and horizontal form, fill and seal.

The vertical form, fill and seal machine forms the package from roll stock flexible film and seals it on a vertical plane after enclosing the product. The method is most applicable to dry free-flowing products such as nuts but may also be applied for liquids, such as juice, or fluids, such as catsup. By gravity the product enters the partially formed package down a tube around which the package has been formed. The product can also be force fed such as by auger or piston.

When gravity is the sole determinant of the rate of fill, product density and rate of fall determine the speed of the equipment during the operation. The machine should not enter any subsequent cycle until the product has completely cleared the sealing area, or the product could interfere with the sealing. Vertical form, fill and seal machines are, with few exceptions, intermittent motion machines requiring completion of one cycle before the following cycle is completed.

Several types of vertical form, fill and seal machines exist. These include the three-side seal unit which forms a pillow-type package and the four-side seal unit often used for liquids. Four-side sealing reduces the possibility of leakage because seals are face-to-face of compatible materials. The pillow-type package with a back seam has problems in the area of the back seam because the seal is formed through a double thickness at the intersections of the top and back seals and, further, may involve attempting to seal two different materials. Dry products may be filled on either type of equipment depending on the volume of the container. The pillow-type of construction achieves greater cubic capacity for each square inch of material than does the four-side pouch.

Stiff packaging materials cannot be formed readily into bag-style construction on vertical form, fill and seal equipment because of the destruction of barrier properties and weakening of the film as it comes over the forming collars and is folded and creased. This operation can fracture individual plys of the web and thus adversely affect barrier properties. These problems have been reduced somewhat through the use of very long folding forms such as on Bemis Company's Ultra-Pak machine which forms a tetrahedral type of package.

## Vertical Form, Fill and Seal

In most vertical machines, the fold is quite severe, and, as a result, heavy laminations of foil, paper and polyethylene tend to crack causing a loss in barrier properties. In contrast, soft films such as polyethylene handle relatively well at this operational stage without loss of properties.

The general form of the pillow pouch is two end seals and a back seal at the middle or along one side. The side seal is not usually found on older American machines, although some more recent American machines reportedly can make side seals. The European Hamac Hansella machine has a side seal so that the graphics may be printed on both front and back of the package.

Generally, however, three-side seal pouches are formed by drawing film from a roll. The film is formed around a collar which is, in turn, around a tube (Figure 4). The back seal is formed by either face to face fin or overlap heat sealing. A horizontal bar then draws the film down the length of one package simultaneously forming the end seal. The bag may then be cut off while the top remains open, or the cut-off station may be at another location. The draw bar then opens and returns to the top of the bag where it forms the top of the bag and the bottom of the following bag. Bag forming is thus an intermittent operation. One face of the draw bar is heated to effect a heat seal on the material.

The distance the film is drawn is regulated principally by the length of the draw bar stroke, which limits the speed of the operation. A photoelectric eye adjusts the motion to a secondary degree so that fully registered bags can be made. During the period the draw bar is being opened and moving to the top of the bag, the product is dropped into the bag, sometimes from a relatively great height. The contents may be forced in at the top but they still must fall through the tube to the bottom. If the heat seal is still tacky, the product can get stuck in the seal area. If the product does

not clear the top seal by the time the draw bar comes across, the product can get stuck in the seal area and prevent effective sealing.

Some vertical form, fill and seal machines employ driven rolls which assist the draw bar in pulling the material as, for example, in some Hayssen equipment. The Rovema machine employs rubber wheels beside the pouch to drive the formed pouch past a rigid fixed sealing and cut-off bar. Intermittent motion machines have rated speeds of 30 to 100 per minute.

Among the vertical form, fill and seal machines used in the United States are those made or sold by Mira-Pak, Inc., Package Machinery Co., Packaging Machinery Division of FMC Corp.; Hayssen Manufacturing Co. (Division of Bemis Corp.); Pneumatic Scale Co.; The Woodman Co., Inc.; Triangle Package Machinery Co., and, of course, many others.

On twin-web machines which fabricate four-side seal pouches, the forming tube is not necessarily circular or oval but rather somewhat diamond shaped (Figure 5). The pouch is pre-shaped prior to product insertion. Materials used in this equipment must be stiff. Because the material does not travel over forming collars it offers few problems relative to stress cracking. If the material is too stiff, however, pouch capacity is reduced and a larger area of packing material is required. Thus, combinations of the paper, foil, and sealant must be carefully selected to assure that there is sufficient cubic capacity to be able to contain the contents.

A number of variations on the conventional vertical form, fill and seal machine are commercially manufactured. These include the Mira-Pak machine in which the filling cup moves down and the pouch moves up to meet the product, thus reducing the drop distance for the product. Mira-Pak is widely used for potato chips. The Rovema machine allows for gusseting along the sides so as to increase the cubic capacity of the formed package. Mira-Pak has a machine which forms a square bottom as well as a gusseted side, thus increasing the cubic capacity of the packaging film considerably and, in effect, making a flexible carton. The tetrahedral package reduces the amount of material required to contain a given quantity of contents, particularly for small size pack-

## FIGURE 4
### Three-Side Seal Pouch Formed on Vertical Form, Fill and Seal Machine

Product

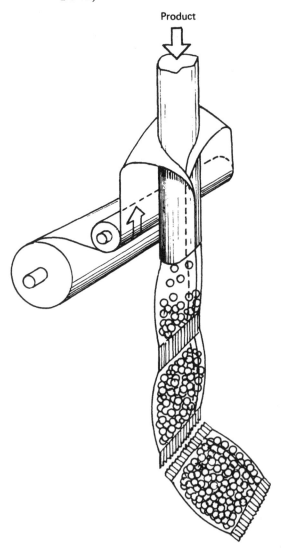

From *Package Engineering*, Jan. 1964. With permission.

ages. Tetrahedral packaging machines are made by two companies: Tetra-Pak of Milliken, S.C., a part of the Swedish firm; and Ultra-Pak, made by Hayssen.

Tetrahedral packages are difficult to align and stack in secondary packages. As a result, tetrahedrons take up considerable volume in a total packaging system. The cost of packaging must be measured in more than just the material used for the primary package.

## FIGURE 5
Four-Side Seal Pouch Formed on Vertical Form, Fill and Seal Machine

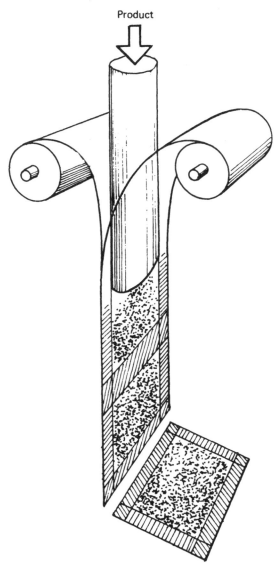

From *Package Engineering*, Jan. 1964. With permission.

The tetrahedral packages' major applications in the United States have been for specialties where specific patterns could be formed and for coffee creamer packages where large numbers of packages are required for very small quantities of contents. Material saving becomes relatively significant when compared against more commonly used thermoform, fill and seal packages which use considerably more material for the contents.

Intermediate between the vertical and horizontal machines are the slant machines typified by the Woodman machine. This machine is similar to a vertical machine except that instead of being vertical, it operates at a slant, thus allowing for a more gentle fall of the contents. As a result, the Woodman machine has found widespread application in filling fragile products such as potato chips and pasta.

### Horizontal Form, Fill and Seal

Horizontal form, fill and seal machines, by definition, form the package while the film is traveling in a horizontal plane. The machine fills the product simultaneously or on a subsequent separate operation from either a horizontal or a vertical direction. This principle may be used either for liquid or dry products because the seals are around four (or three) edges. Any free flowing product that is filled on a horizontal form, fill and seal machine is generally filled from the vertical axis so that it can contain the product until the machine effects the final seal on the pouch.

The major horizontal form, fill and seal machine in the United States is the Bartelt, used principally for dry powdery materials which must be protected against the external environment. Because the material is not damaged by machining, high barrier films can be used and sealed to protect the product against the environment. The principal Bartelt competition are the Canadian Delamere and Williams, which runs on much the same principle, and several pieces of European equipment such as the Höller which forms the pouch at one station and fills and seals on a rotary turret at another station.

Bartelt equipment can form and fill the package on intermittent motion equipment, which can run from 60 to 120 per minute or on continuous motion equipment which can make up to 300 packages per minute by forming several pouches simultaneously on a continuous conveyor. Continuous motion machines require multiple filling heads, whether for liquid or for powders, so that several packages are filled simultaneously.

Intermittent motion machines can effectively utilize multiple filling heads but often employ a single filling head.

The horizontal form, fill and seal machine is used for such products as drink mixes, powdered soups, and other dry foods which require good protection against moisture.

Packages made on the horizontal form, fill and seal machine may be made from single or double webs but most often the single web is used on Bartelt-type equipment (Figure 6). This single web is folded, and a bottom and two side seals are formed simultaneously. Fin seals allow sealant to be coated or extruded on the inside. When aluminum foil is employed in such a combination, the foil must be protected by a material, such as paper, on the outside. Paper prints better than does the aluminum foil. More important, however, the paper protects the aluminum foil against physical damage and abuse from external scratching and pinholing. The materials used may, of course, be stiff. The stiffer they are, the less the volume of contents.

Contents may be filled under a controlled atmosphere, e.g., an inert gas such as nitrogen or carbon dioxide. Horizontal form, fill and seal equipment lends itself better to controlled atmosphere than does the vertical form, fill and seal, although some vertical machines are now being employed for gas flush packaging for such products as ground coffee. In general, on vertical machines, the gas flush occurs during the formation of the package. On occasion, however, such as on the Rovema or FMC Corp.'s Stokeswrap, the pouch is formed and filled, left open at the top, and enters a chamber at another station for gas flushing or vacuumization.

Horizontal form, fill and seal machines may also use the film in a horizontal plane. The product can enter in a vertical plane, between two webs of film in which the product is inserted from the side, or from the top with a second web coming down on the top to form a four-side seal around the package (Figure 7). An example is the Rotowrap machine.

FIGURE 6
Four-Side Seal Pouch—Vertical Feed Schematic Diagram
Typical of Bartelt Equipment

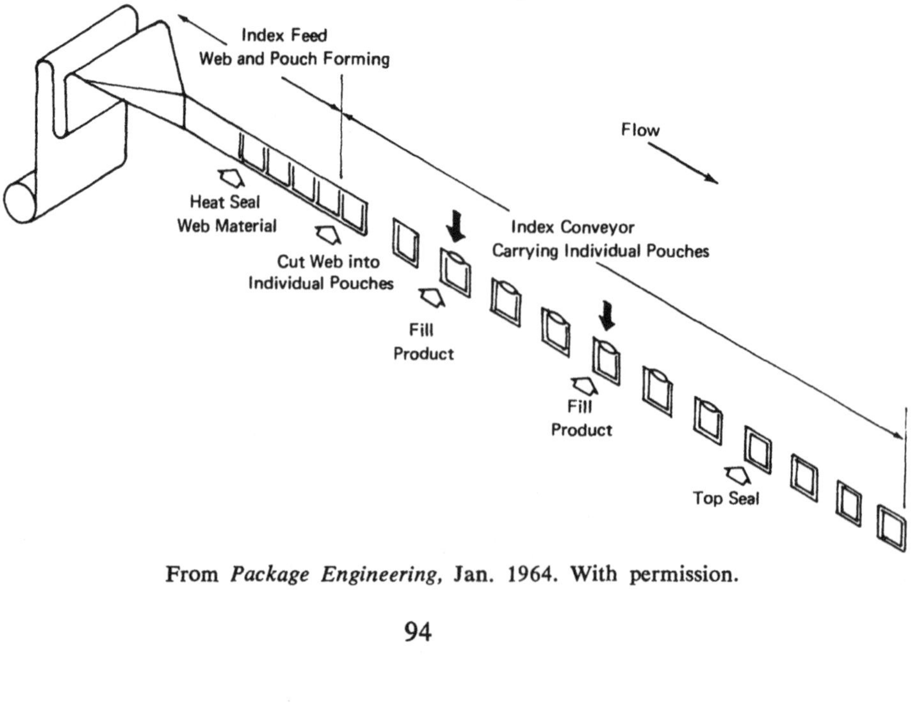

From *Package Engineering*, Jan. 1964. With permission.

### Other Systems

In addition to these major types, a number of specialty types exist. An example is a vertical form, fill and seal using two sheets of film to fill liquids or tablets as on a Crompton and Knowles unit. Relatively stiff packaging material can be used because the sealant is welded on the inside. A continuous motion vertical machine for liquids is the Circle equipment.

One application of a horizontal form, fill and seal is the Hayssen RT machine in which a single web of film is formed into three-side fin seal pouch around products in continuous motion with a simultaneous gas flush. This unit (Figure 8) is employed for such products as cured meats and cheese. The machine operates

#### FIGURE 7
Four-Side Seal Pouch with all Webs Formed from Horizontal Plane

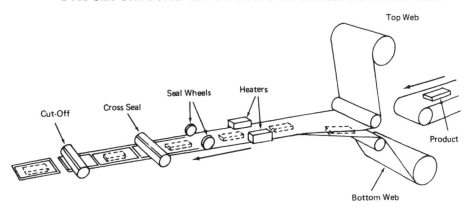

From *Package Engineering*, Jan. 1964. With permission.

#### FIGURE 8
Hayssen RT Equipment

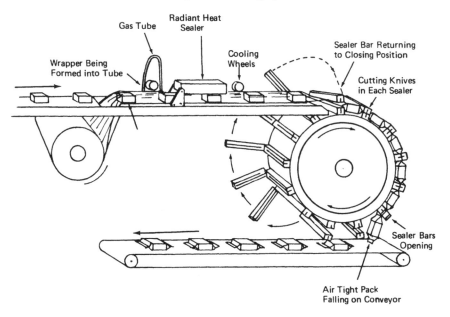

From American Can Co., Greenwich, Conn. With permission.

on a continuous motion; and, as a result, relatively high speeds can be achieved. Because there is no forming head, the film is not stressed. The back seal is formed above the product and the end seals are formed on a rotating sliding die drum which allows for a variation in the cut-off length as well as in the positioning of the seal.

The cookie, cracker and candy industries use what are essentially horizontal form, fill and seal machines forming packages from roll stock. Examples of this unit are the Hudson-Sharp Campbell wrapper, Doughboy machine, and SIG HSU machines (Figure 9). Roll stock is drawn from the roll in a continuous motion and pulled around the product which is generally small but can be as long as 12 inches. An overlap or face-to-face seal may be formed on the bottom, but there is no forming horn around which the film can stress or crack. This operation can be performed at very high speed because of continuous motion. The end seal is formed by a rotating head seal anvil which, unfortunately, must effect a seal within a very brief time, simultaneously cutting the film. During this very short time, cutting must occur without tearing. The takeaway belt must be faster than the packaging element to separate the packages. This type of packaging machine is used for portion packaging of crackers and

for individual wrapping of candy bars. Because end fins can extend out, hold-under equipment for the seals is sometimes added; but this modification leads to intermittent motion. Points of fracture can occur at the corners where the film stresses around the corners of the product.

A modification of the horizontal feed, vertical fill machine enjoys widespread use for processed meat and frozen boil-in-bag food packaging. One web, usually of polyester/polyethylene or nylon/polyethylene, is vacuum formed into a cavity. The product is manually (or sometimes automatically) inserted (Figure 10). A second film web (usually of a less expensive material because it does not require thermoforming) is four-side sealed to the formed web. A small opening is left to allow for vacuumization or gas flushing. The final seal is made after the new atmosphere is incorporated.

Both Standard Packaging Corp. and Mahaffay and Harder Engineering Co. manufacture this type of equipment.

The double package maker, although not strictly a piece of flexible packaging equipment, includes flexible material, usually waxed glassine or glassine lamination. A glue or wax-sealed bag-like structure is formed over a mandrel from roll stock material. Wrapped around the flexible bag is a paperboard carton made

FIGURE 9
Pouch Style Wrapper

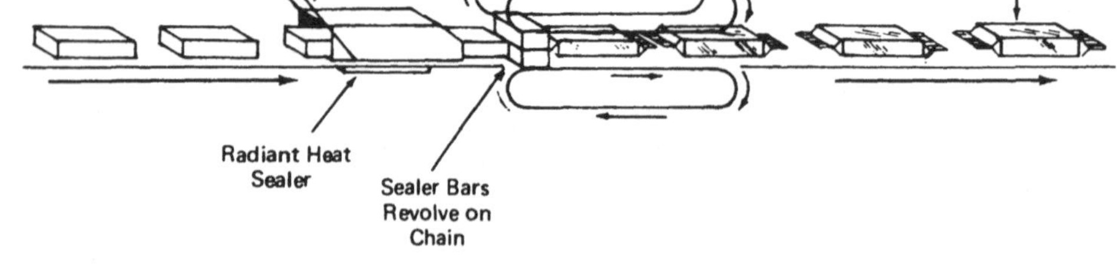

Schematic diagram typical of Hudson-Sharp Campbell Wrapper. From American Can Co., Greenwich, Conn. With permission.

either from a flat blank or a preformed carton. Combining is performed on a multiturret machine with one operation being done at each turret. The combined open bag and carton is discharged from the mandrel to a conveyor which transfers it to a filling station. Closure after filling may be foldover, glue-seal or wax-seal. Operating speeds of up to 100 per minute are common.

In the past few years, several European companies have introduced a new form of packaging machine, the thermoform, fill and seal machine. A number of these have been in use in the United States for many years, but these were designed and built by the packagers. Being high speed, they are often used for portion packaging for such products as jams, jellies, condiments, and refrigerated coffee cream.

Such machines essentially utilize roll stock thermoforming sheets such as PVC. The sheet is heated much as a blister forming sheet would be heated, and the sheet is vacuum-formed in a die (Figure 11).

## FIGURE 10
### Vacuum Forming Machine used for Processed Meats

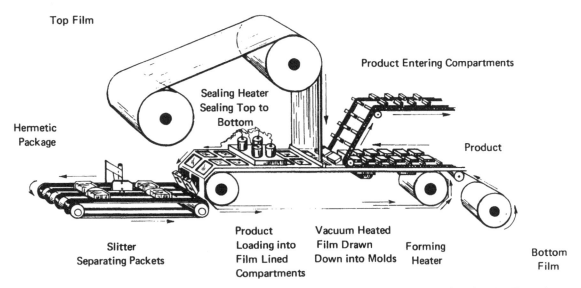

Typical of equipment made by Standard Packaging Corp. or Mahaffey and Harder Engineering Co. From American Can Co., Greenwich, Conn. With permission.

## FIGURE 11
### Thermoform, Fill and Seal Equipment Typical of
### Anderson Bros. Mfg. Co.

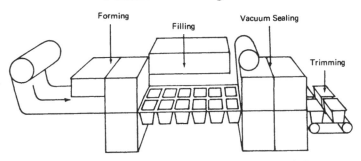

From *Package Engineering*, Jan. 1969. With permission.

A filling head seats itself over the top of the cavity and fills it with fluid or semi-fluid contents. After completion of filling, a compatible thermoplastic film or film laminate is brought down over the top of the container and sealed on flat flanges formed between the cavities. After heat sealing, a large cutting die separates the sheets or individual packages. Generally, this operation is performed in multiples so that large numbers of small packages can be formed simultaneously.

The packaging material, of course, must be thermoformable with no residual odor being imparted to the contents as a result of thermoforming. Further, because the material is still warm when the contents are inserted, there should be no interaction between the plastic and the contents. The material should be readily formable because there are no safety mechanisms for separating the improperly formed from the properly formed containers.

Although the number of cycles per minute is relatively small, large numbers of packages are produced because of the several-across aspect. There must be provision for taking away large numbers of these containers which are formed simultaneously. This is often accomplished by dump filling into secondary containers.

Although the commercial machines employed are generally of European origin, several of the more popular are being marketed by firms such as Anderson Brothers Mfg. Co.

## CONCLUSIONS

One conclusion emerges with stark clarity: there can be no critical review of flexible packaging in the food industry at present because there are too few published facts to review. This field is vast and broad because it grew by empirical means without information. Special interest groups promoted what suited their purposes, and gained profit as the result. An intricate network of basic material suppliers and converters still hustles to compete for the packager's dollar.

Scientists work in packaging supplier laboratories developing special materials. They then learn how to treat them without having any no-tion of the uses that might be made by their findings. Technologists in packaging supplier organizations are few and far between, and are generally stretched thin.

Marketing and technical men (among packagers) seek and ask for packaging to meet the ever-changing packaging requirements, but somehow this communication does not get through to the supplier. A dichotomy of packaging without uses and needs without solutions has developed.

The most critical comment that can be made about the technical literature is that it covers the properties of materials very well. Documentation of film properties is excellent.

This is next to useless for a user who is performance oriented. It tells him nothing about what kind of package can be made and what the characteristics of the package might be. Furthermore, the data are generally colored by the interests of the author's company: drawbacks and weaknesses of the material are not highlighted.

A glance at the reference material on actual usage of flexible materials for foods reveals one fact: scarcity. A single author has published more than all of the named authors combined. The best articles are anonymously written as editorial material.

If these are harsh judgments, then there must be some other explanation for the absence of data. Gross differences in packaging materials used for the same products by different companies, both with the explanation that testing has demonstrated the superiority of one material over the other, must be explained. This obvious disparity looms largest. The multitude of materials with relatively few actually being used must be explained. There must be reasons why small converters, with no technical capabilities, can provide highly specialized materials, while at the same time large scientifically oriented converters continue with declining production of old-line flexible structures.

Objective packaging audits often disclose the deficiencies and miscalculations of packaging. Most food companies have no packaging personnel. Those that do often pigeonhole them into specialties such as glass or testing or graphics. Gross inadequacies in knowledge beyond specialties exist. Apparent solutions, generally offered by suppliers with little knowledge of packagers' problems, are accepted **without question.**

This review of flexible packaging in foods has attempted to underscore the void between packaging requirements and the fulfillment of these requirements. Certainly two rationales for any type of packaging are cost and performance tests, and this situation will continue as long as equipment, materials, and user people have no dialogue. This situation will also continue as long as the universities focus on minute, precisely definable, quantitative problems instead of playing on the whole spectrum. It will continue until the sales expert, whose only motive is self-profit, is displaced by the dedicated with insight. Until managements acknowledge that packaging is as critical to their organization's operation as any other phase of the business we will face the same problems. Needless to say, today's confusion is the opportunity for tomorrow's rationalization.

## REFERENCES

1. Allen, Nelson, The evolution of packaged meats, *Meat*, 35, 23, 1969.

2. Angel, T.H., Basic properties of aluminum foil, *Packaging*, 39, No. 459, 36, 1968.

3. Anon., Cans: challenge to metal, *World Coffee & Tea*, 8, No. 8, 39, 1967.

4. Anon., Close-up: meat, *Package Engineering*, 12, No. 12, 83, 1967.

5. Anon., *The Commercial Storage of Fruits, Vegetables, and Florist and Nursery Stocks*, U.S. Department of Agriculture, Agricultural Handbook No. 66, October, 1968.

6. Anon., Evolution into diversity, *World Coffee & Tea*, 8, No. 8, 11, 1967.

7. Anon., Film lids on bake-in-oven frozen vegetable casseroles, *Modern Packaging*, 42, No. 3, 86, 1969.

8. Anon., Flexible packaging: confused and complicated, *Modern Packaging*, 40, No. 14, 122, 1967.

9. Anon., Flexible packaging: the common denominators, *Modern Packaging*, 40, No. 15, 136, 1967.

10. Anon., Flexible packaging: the overlooked components, *Modern Packaging*, 40, No. 16, 112, 1967.

11. Anon., Flexible packaging for pasteurized foods, *Food Engineering*, 38, No. 7, 76, 1967.

12. Anon., Gas flushes coffee's oxygen, tough pouch keeps it out, *Package Engineering*, 13, No. 12, 77, 1968.

13. Anon., Machines for non-atmospherics, *World Coffee & Tea*, 8, No. 8, 30, 1967.

14. Anon., Materials: where to put the midpoint? *World Coffee & Tea*, 8, No. 8, 36, 1967.

15. Anon., Packaging tackles the cold and the dry, *Package Engineering*, 13, No. 11, 87, 1968.

16. Anon., Preportioning: leverage for less water? *World Coffee & Tea*, 8, No. 8, 14, 1967.

17. Anon., Revolution in institutional packs, *World Coffee & Tea*, 8, No. 8, 22, 1967.

18. Anon., 'Revolutionary' packaging mainstay of the frozen food industry, *Quick Frozen Foods*, 30, No. 12, 1968.

19. Anon., *The Search for a Thousand Million Dollars*, A.T. Kearney & Co., Inc., Chicago, 1966.

20. Anon., Snack packaging past and present, *Snack Food*, 58, No. 3, 34, 1969.

21. Anon., Spin-off into home flexibles, *World Coffee & Tea*, 8, No. 8, 27, 1967.

22. Anon., Surge into non-atmospherics, *World Coffee & Tea*, 8, No. 8, 18, 1967.

23. Anon., Understanding the costs of flexible packaging, *Snack Food*, 58, No. 3, 42, 1969.

24. Anon., What's new in dairy products packaging? *Canadian Dairy & Ice Cream Journal*, 47, No. 6, 29, 1968.

25. Atkin, Lawrence, Workshop on flexpack ready-to-eat foods, *Transcripts of the Workshop Sessions on Military Food and Container Problems,* October 24, 1967.

26. Brody, A.L., Shelf life parameters - meat, *Modern Packaging,* in press.

27. Brown, Marion L., Gas packaging for chips, snacks, *Snack Food*, 58, No. 3, 44, 1969.

28. Cook, Russell J., How the package manufacturer helps design packages for profit, *Proceedings of the Seminar on the Packaging of Poultry: Fresh and Frozen*, New Orleans, March 5, 1969.

29. Eichhorn, Jacob, Films, *Mosher & Davis/Industrial & Specialty Papers*, 11, 246, 1968.

30. Hannan, R.S., Some properties of food packaging materials which relate to the microbial flora of the contents, *Journal of Applied Bacteriology*, 25, No. 2, 248, 1962.

31. Holland, J. Kent, Institutional coffee: vacuum pack or gas flush? *World Coffee & Tea*, 8, No. 8, 43, 1967.

32. Howard, Anne, Sausages strive for shelf appeal, *Packaging Review*, 88, No. 12, 28, 1968.

33. Ingram, M., Microbiological principles in prepackaging meats, *Journal of Applied Bacteriology*, 25, No. 2, 259, 1962.

34. Karitas, James, J., *Packaging Produce in Trays at the Central Warehouse*, U.S. Department of Agriculture, Marketing Research Report, No. 827, 1967.

35. Lawrie, R.A., *Meat Science*, Pergamon Press, Oxford, 1966.

36. Manne, Stanley, (Abstracted Article From Address), Package change dynamics, *Meat*, 35, No. 7, 29, 1969.

37. Martin, Edward L., The use of foil laminates, *Food Packaging with Flexible Laminates - Materials, Machines, Quality Control, Marketing*, Proceedings of a University of California Packaging Program Seminar, November, 1967.

38. Mayer, Peter C. and Robe, Karl, Progress report on - canning without cans, *Food Processing*, 24, No. 11, 77, 1963.

39. Rubinate, Frank J., Army's 'obstacle course' yields a new look in food packaging, *Food Technology*, 18, No. 11, 71 1964.

40. Sacharow, Stanley, Boil-in-bag market, *Paper, Film and Foil Converter*, 40, No. 3, 47, 1966.

41. Sacharow, Stanley, Cheese...A. 1.5 billion lb. packaging market, *Paper, Film and Foil Converter*, 40, No. 6, 45, 1966.

42. Sacharow, Stanley, Confectionery packs demand attention, *Paper, Film and Foil Converter*, 40, No. 11, 93, 1966.

43. Sacharow, Stanley, Convenience pack market demands converter know-how, *Paper, Film and Foil Converter*, 40, No. 9, 63, 1966.

44. Sacharow, Stanley, Deep-freeze storage is tough test for converted packages, *Paper, Film and Foil Converter*, 41, No. 1, 54, 1967.

45. Sacharow, Stanley, Fresh produce - growing market for flex packs, *Paper, Film and Foil Converter,* 41, No. 7, 41, 1967.

46. Sacharow, Stanley, Foil and snack food: a happy marriage, *Snack Food*, 58, No. 3, 39, 1969.

47. Sacharow, Stanley, Frozen food packaging, *Food in Canada*, 28, No. 12, 33, 1968.

48. Sacharow, Stanley, Laminates move to wrap up huge processed meat market, *Paper, Film and Foil Converter*, 40, No. 12, 60, 1966.

49. Sacharow, Stanley, Market potential perks up for coffee flex packs, *Paper, Film and Foil Converter*, 41, No. 6, 41, 1967.

50. Sacharow, Stanley, Poly bags seize bread market, *Paper, Film and Foil Converter*, 41, No. 3, 39, 1967.

51. Sacharow, Stanley, Prepacks take over poultry market, *Paper, Film and Foil Converter*, 41, No. 4, 102, 1967.

52. Sacharow, Stanley, Snack packs - where the action is in new laminations, *Paper, Film and Foil Converter*, 40, No. 7, 62, 1966.

53. Sacharow, Stanley, Special machinery adapts plastic film for use, *Candy Industry*, 131, No. 6, 1968.

54. Schertz, E.P. and Brody, A.L., *New Developments in Meat and Meat Packaging Technology*, Report for Iowa Development Commission, Arthur D. Little, Inc., Cambridge, Mass., 1968.

55. Shaffer, Paul F., *Packaging Produce at the Central Warehouse*, U.S. Department of Agriculture, Marketing Research Report No. 721, 1966.

56. Simms, William C., Ed., *Modern Packaging Encyclopedia, 1969*, McGraw - Hill Co., New York, 1969.

57. Southam, E.V., Flexible laminates, *Packaging Technology*, 15, No. 104, 17, 1969.

58. Tomkins, R.G., The conditions produced in film packages by fresh fruits and vegetables and the effect of the conditions on storage life, *Journal of Applied Bacteriology*, 25, No. 2, 290, 1962.

59. Vass, David, Mcd., Saran: high barrier coating for snack packaging, *Snack Food*, 58, No. 3, 38, 1969.

60. Williams, G.F., How to select a flexible aluminum foil packaging laminate, *Packaging*, 39, No. 459, 46, 1968.

## ADDITIONAL BIBLIOGRAPHY

61. Allen, Nelson, Films in the 70's, American Management Association 38th National Packaging Conference, Chicago, April 16, 1969.

62. Almarker, Carl A. and Wallenberg, K. Eric, S., New pouch method for ground coffee, *Modern Packaging*, 41, No. 6, 146, 1968.

63. Anon., Coextruded films seen as boost for profitable package design, *Canadian Packaging*, 22, No. 3, 36, 1969.

64. Anon., *Plastics in Packaging*, Welwyn Gorden City, Herts, England, ICI, 1965.

65. Anon., Polyflex bi-axially oriented polystyrene sheet - recent developments, *Packaging*, 40, 62, 1969.

66. Anon., Vacuum packed steaks being shipped unfrozen by Denver purveyor, *Western Meat Industry*, 1965.

67. Anon., Vacuum sealing portion cuts in portion packs, *Western Meat Industry*, 1966.

68. Anon., 'Valeron' high-density polyethylene film, *Packaging*, 39, No. 465, 45, 1968.

69. Anon., What's ahead in form - fill - seal packaging, *Package Engineering*, 14, No. 1, 1969.

70. Bengtsson, O. and Bengtsson, N.E., Freeze-drying of raw beef III. Influence of some packaging and some storage variables, *Journal of the Science of Food and Agriculture*, 19, No. 9, 486, 1968.

71. Burton, D.R., Transparent foil combinations, *Food Packaging with Flexible Laminates*, Proceedings of University of California Packaging Program Seminar, November, 1967.

72. Cage, James K., Developing and effecting quality standards, *Food Packaging with Flexible Laminates*, Proceedings of University of California Packaging Program Seminar, November, 1967.

73. Clemen, J.D., Vacuum packaging machinery, *Food Packaging with Flexible Laminates*, Proceedings of University of California Packaging Program Seminar, November 1967.

74. Davis, E.G., and Burns, R.A., Oxygen permeability of flexible film packages for foods, *Food Technology*, 23, 92, 1969.

75. De Majc, W.M., Designing aluminum foil packs, *Packaging,* 39, No. 459, 42, 1968.

76. Eheart, Marys, Variety, fresh storage, blanching solution and packaging effects on ascorbic acid, total acids, pH and chlorophylls in broccoli, *Food Technology,* 23, 238, 1969.

77. Eichhorn, J., A series of new barrier films, Dow Chemical Company, 1967.

78. Fellers, David A., Wahba, Isaac, J., Caldano, J.C., and Ball, C. Olin, Factors affecting the color of packaged retail beef cuts - origin of cuts, package type and storage conditions, *Food Technology,* 17, 1175, 1963.

79. Gilbert, Seymour G., and Pegaz, David, Find new way to measure gas permeability, *Package Engineering,* 14, No. 1, 66, 1969.

80. Goff, James W., Brooks, Hallard N., and Hendee, John R., *A Study of the Potential Demand for In-Plant Package Manufacturing Machinery,* Michigan State University, January 1, 1965.

81. Gordy, T.L. and Kresse, H.J., Methods of coating flexible substrates with hot melts, *New Developments in Wax Technology for Packaging,* Proceedings, University of California Seminar, October 19, 1967.

82. Guillottee, John E., Coextrusion gets workhorse PE film into some new competition, *Modern Plastics,* 46, No. 1, 1969.

83. Healy, Gene, Atmospheric machines - a diverse variety of types developed to the precise needs of the product, *Food Packaging with Flexible Laminates,* Proceedings of University of California Packaging Program Seminar, November, 1967.

84. Heyer, Kai L. and Raup, R.S., Now on the film scene: saran-coated polypropylene, *Package Engineering,* 14, No. 3, 63, 1969.

85. Highland, H.A., Guy, R.H., and Laudani, H., Polypropylene vs. insect infestations, *Modern Packaging,* 41, No. 12, 113, 1968.

86. Highland, Henry A., Secrease, Margaret, and Merritt, P.H., Polyvinylidene - coated Kraft paper as an insecticide barrier in insect resistant packages for food, *Journal of Economic Entomology,* October, 1968.

87. Hiley, Alan E., Thermoplastic films offer great versatility for candy packaging, *Candy Industry,* 132, No. 11, 41, 1969.

88. Karel, M., Issenberg, P., Ronsivalli, L., and Jurin, V., Application of gas chromatography to the measurement of gas permeability of packaging materials, *Food Technology,* 17, 327, 1963.

89. Lampi, R.A., Resistance of flexible packaging materials to penetration by microbial agents, U.S. Department of Commerce, National Bureau of Standards, Institute for Applied Technology, AD 651493, April, 1967.

90. Leonard, Edmund A., *The Economics of Packaging,* New York: Packaging Institute, 1968.

91. Lynch, L.W., Evaluations of plastic films, *Modern Packaging,* 37, No. 10, 177, 1964.

92. McCarron, Robert M., Cellophane can be tailored to a variety of candy packaging requirements, *Candy Industry,* 132, No. 11, 1969.

93. Maunder, D.T., Folinazzo, J.F., and Killoran, J.J., Bio-test method for determining integrity of flexible packages of shelf stable foods, *Food Technology,* 22, 615, 1968.

94. Mohaupt, A.A., Tests give film barriers the air, *Package Engineering,* 13, No. 12, 80, 1968.

95. Osmon, F.O. and Wolff, V.C., Surlyn® D ionomer dispersions - chemistry, functional coatings and summary of applications, *TAPPI,* September 19-21, 1966.

96. Paine, F.A., *Packaging Materials and Containers,* Blackie and Sons, Limited, London; Glasgow, 1967.

97. Paine, F.A. and Riddell, G.L., Shelf life estimation of moisture sensitive products - a new approach, *VERPAC-KUNGS RUNDSCHAU,* 19, No. 9, 1064, 1968.

98. Paine, F.A. and Stocker, J.H.J., Heat-sealing foil packs, *Packaging,* 39, No. 460, 69, 1968.

99. Pierson, M.D. and Ordal, Z.J., Microbiological, sensory, and pigment changes in aerobically and anaerobically packaged beef, Presented at IFT Meeting, Chicago, May 15, 1969.

100. Sacharow, S., Methods of identifying packaging materials are outlined, *Candy Industry*, 132, No. 1, 1969.

101. Sacharow, S., Packaging labels often contain a number of hidden costs, *Candy Industry*, 132, No. 10, 1969.

102. Sacharow, S., Permeability of package is prime determinant of shelf life, *Candy Industry*, 131, No. 12, 1968.

103. Scott, Charles R., Product-oriented food packaging, Presented at Highlights of Food Science Conference, Michigan State University, March 17, 1969.

104. Shank, J.L. and Lundquist, B.R., The effect of packaging conditions on the bacteriology, color and flavor of table-ready meats, *Food Technology*, 17, 1163, 1963.

105. Shaw, Fred B., You can have your seal, yet open it by peeling, *Package Engineering*, 14, No. 4, 93, 1969.

106. Short, Donald, Latent lures in laminates, *Food Packaging with Flexible Laminates*, Proceedings of University of California Packaging Program Seminar, November, 1967.

107. Shorten, D.W., Choosing the right film for the job, *Packaging Technology*, 14, No. 102, 17, 1968.

108. Taylor, A.A., Initial gas exchanges in the $CO_2$ packing of bacon. Gas exchanges in packaged meats 2, *Food Processing and Packaging*, 25, No. 7, 1964.

109. Wilks, Robert A. and Gilbert, Seymour, Measurements of volatiles transferred from plastic packaging films to foods, *Food Technology*, 23, 47, 1969.

110. Anon., *Flexible Packaging in Maraflex*, American Can Co., Neenah, Wisc., 1963.

111. Anon., *1970 Package Machinery Directory*, Packaging Machinery Manufacturers Institute, Washington, 1970.

112. Anon., Packaging Machinery Catalog Issue, *Package Engineering*, 1970 Edition.

113. McGillan, Donald and Neacy, Thomas, Choosing machinery for flexible packaging Part I. Vertical and horizontal form, fill and seal, controlled atmosphere, and preformed bags and pouches, *Package Engineering*, 9, No. 1, 1964.

114. McGillan, Donald and Neacy, Thomas, Choosing machinery for flexible packaging, Part II. Overwrapping, direct or intimate wrapping and vacuum form, fill and seal, *Package Engineering*, 9, No. 2, 1964.

115. McGillan, Donald and Neacy, Thomas, Choosing machinery for flexible packaging, Part III. Conclusion, package closing, heat sealing, and choosing the right machine, *Package Engineering*, 9, No. 3, 1964.

116. Sacharow, Stanley and Griffin, Roger, *Food Packaging*, Avi Pub. Co., Westport, Conn., 1970.

117. Schertz, E.P. and Brody, A.L., *Convenience Foods: Products/Packaging/Markets*, Report for Iowa Development Commission, Arthur D. Little, Inc., Cambridge, Mass., 1970.

T - #0714 - 101024 - C0 - 266/188/6 - PB - 9781138558731 - Gloss Lamination